WHY THE CORPORATE WORLD SUCKS

AND THE SECRET LESSONS TO SUCCESS THEY DON'T WANT YOU TO KNOW

Wayne Vogel

This book is dedicated to: Anyone who has ever felt the pain of being overlooked professionally, passed on for a deserved promotion, or taken advantage of for the sheer greed of others, I wrote this book for you. Also, to my good friend Richie, who referred me to my first IT job, which started all of this—RIP, my brother.

TABLE OF CONTENTS

Table of Contents..........iv

Preface..........1

Section 1 – Introduction..........4

Reason One: Everyone Always Complains..........5

Reason Two: Most Executives Are Not Qualified..........7

Reason Three: Certain Employees ALWAYS Know It ALL..........11

Reason Four: Senior Management Staff Are Terrible Decision Makers....14

Section 2 – Systemic and Foundational Challenges..........17

Reason Five: Corporations Move Slowly (Except When They Change).....18

Reason Six: Corporations Try to Change Too Fast..........22

Reason Seven: Corporations Fail on Large Programs and Projects..........25

Reason Eight: Corporations Don't Understand Data..........28

Reason Nine: They Don't Understand Processes..........34

Reason Ten: They Don't Know How to Automate..........37

Section 3 – Day-to-Day Frustrations..........40

Reason Eleven: Corporations Utilize All of Your Flow..........41

Reason Twelve: Corporations Stifle Your Ability to Learn..........44

Reason Thirteen: Corporate Leaders Always Want to Run Their Mouths..47

Reason Fourteen: Corporations Force Pointless Meetings..........49

Reason Fifteen: Corporations Have Too Many Rules and Don't Adapt....52

Section 4 – Cultural and Leadership Issues..........55

Reason Sixteen: Corporate Leaders Don't Respect You (Or Themselves).56

Reason Seventeen: Corporations Cultivate a Superstar Mentality..........59

Reason Eighteen: Corporations Breed a Culture of Multitasking..........62

Reason Nineteen: Corporations Are Not Trustworthy……………………..65

Reason Twenty: Corporate Leaders Play Favorites………………..………...68

Section 5 – Operational Inefficiencies……………………………………..……...71

Reason Twenty-One: Corporate Leaders Don't Challenge Vendors or Partners Enough……………...……………………………………………………..…...72

Reason Twenty-Two: Corporate Leaders Overcommit on Your Behalf….76

Reason Twenty-Three: Corporations Don't Know How to Sell……..…....79

Reason Twenty-Four: Corporate Leaders Overcomplicate the Simple and Oversimplify the Complex……………………………………….……………...83

Reason Twenty-Five: Corporations Don't Understand Roles and Responsibilities………………………………………………………………..…86

Reason Twenty-Six: Corporations Don't Understand Planning…….....….90

Section 6 – Business-Specific Challenges…………………………….………….94

Reason Twenty-Seven: Corporations Take Business Too Personally…....95

Reason Twenty-Eight: Corporations Don't Really Understand Business…98

Reason Twenty-Nine: Corporations Get Excited About Business They Don't Yet Have……………………………………………………………………………101

Reason Thirty: Corporations Get Excited About Business They Don't Understand……..……………………………………………………………….104

Reason Thirty-One: Corporate Leaders Don't Understand Staff……..….107

Section 7 – Philosophical and Big-Picture Reflections……………………..…...112

Reason Thirty-Two: Corporations Overwork You…...…………………113

Reason Thirty-Three: Some Corporate Leaders and Employees Try to Manipulate You…………………………..…………………………………..117

Reason Thirty-Four: Corporations Take Advantage of You…...………...120

Reason Thirty-Five: Corporate Leaders Claim Their Business Is 'So Unique' or 'So Specialized'…………………………………………………..…………122

Reason Thirty-Six: Corporations Promote the Retirement Carrot……....127

Reason Thirty-Seven: Corporations Require and Then Don't Utilize Higher Education……………………………………………………..…………………………130

Section 8 – Strategic Decisions………………………………………………………...133

Reason Thirty-Eight: Executives Steal Your Ideas……...…………………..134

Reason Thirty-Nine: Corporations Never Have Budget (Until They Do).137

Reason Forty: Corporate Leaders Create Useless Work………….....…….141

Reason Forty-One: Corporations Aren't Entrepreneurial……….……….144

Reason Forty-Two: Corporations Are Always Reactive……..……………147

Conclusion…………………………..……………………………………...150

Bibliography………………...………………………………………………152

Acknowledgments…………….……………………………………………...156

Disclaimer…………………………..……………………………………….158

About the Author…………………………...………………………………159

Preface

The corporate world does suck, doesn't it? I mean, I could just end the book right there and not dig any deeper. I can't, though. Too much of my career was spent dragging myself into buildings I had to be in. Many times, the gut-wrenching feelings were so palpable I contemplated if I was getting ill or not. I am willing to guess you've experienced this oh-so-wonderful feeling a time or two yourself. Where do these feelings come from, though? Are we all just like Peter in the movie *Office Space* who wants to sit around and do nothing all day? In a word, no. The truth about it all will leave you in a state of shock.

Within the contents of this book, you will find the correlations between those feelings and the reasons why we get so many of them. More importantly, because we cannot simply complain all day to feel better, there will be lessons that will make it all make sense. Along the way, you will begin to see the real reasons why these things happen. And dare I say, this collection of reasons and lessons will become some of the most valuable business information you may ever consume. It will help you not only with your career but also with your life.

Before the creation of this book, I read more business and data books than I care to admit. Out of the gate, let's get one thing straight: This book is not an elongated resume. Business books like that should be written with disappearing ink. The best books I read shared that sentiment, so I am following suit. When I set out on this project, my goal was to create one of the most comprehensive corporate and white-collar business books ever written, tackling the *real problems* I saw from living in it for 20 years as an IT engineer, senior manager, and database and business process architect. No matter how much business knowledge a person has, I have not seen or met a business professional who didn't run into at least some of these issues. When I do tell a personal story, it will point to just how much corruption there is.

Once I acquired all of this information, so many truths became clear. Finding the truth in such complicated issues is a value that is difficult to articulate. The depth of which these issues span involves human nature, psychology, emotion, politics, and downright greed. Setting yourself free from the burdens of darkness in these areas feels immensely rewarding. Pulling the curtain to expose the 'masters' behind the strings of us worker puppets tells a story many won't believe or may simply deny altogether. The neat thing, though? You cannot outrun the results and facts of effectiveness. Understanding these truths can make your experience in the corporate world truly effective and save you immeasurable time from worry and angst.

This book also has a 'choose-your-own-adventure' component inspired by my favorite childhood books. Shout-out to the eighty-year-old R.L. Stine, author of the Goosebumps series! The fork in the road here? Whether you will take the scenic route or a shortcut route. I have designed this book with all of the context at the beginning and middle of each chapter. The tips, guidelines, and principles are at the end of each chapter, listed in bullet points for quick consumption and re-referencing. The path you take is up to you.

So, armed with this context in mind, we will journey together into what is sure to be a beneficial understanding and newfound lens through which to view the corporate world. Let's get going and make sense of what really happens when summing up your experiences in the corporate world!

Section 1 - Introduction

Reason One: Everyone Always Complains

I felt it was most appropriate to start here because the irony is evident. The very actions so many of the employees would take earlier in their tenure would result in the very situation they were complaining about at that moment. The lack of logic in their complaining matched that of the hasty overgeneralizations you see all over the internet these days. There's little thought investment in it. The reason for this is clear: They don't truly believe in what they are doing for the company, resulting in venting. Maybe that comes from the lack of control they perceive to have in their positions. Or maybe, they can see through the copious amounts of bullshit being spewed from every direction and don't know how to proceed…

So many times, people find problems just to complain about them. So much of this is just human nature, though, right? Not exactly in this instance. Part of what you are paid for in corporate jobs is to be able to understand where problems within the company are coming from. It's too bad most of them do not address this upfront and pretend everything is wonderful from the moment you walk in. I have never seen a mission statement for all the companies I have worked for state this important initiative: Train your brain to think of problems as opportunities.

Learn to replace all your complaints with the idea that the situation is an opportunity for improvement. Finding and fixing problems can be the difference between keeping your job and losing it. Your coworkers and superiors will always remember who made their lives easier, a benefit that almost certainly comes in handy down the line during tumultuous times. Mastering this skill turns all the seemingly innocuous inaction of complaining into action-filled results. And we all know how much the bosses love results.

In addition to the aforementioned skill, learning to build your understanding of the corporate world will unquestionably deflate your need or desire to complain. At the end of the day, complaint-worthy times will occur. Succinctly put:

> – Learn to resolve the feelings associated with the bad times and remember the lessons from them.
>
> – Remember to train your brain to think of problems as opportunities.

In the next reason, we'll discuss one of the primary reasons employees complain. Let's just say human nature may be spot on with this one.

Reason Two: Most Executives Are Not Qualified

Yes, you read this reason correctly. This one will be a shock to some and obvious to others. While some of this is circumstantial, I've found as a rule that regardless of how intelligent, articulate, and downright kind some of these individuals are or appear to be, they do not have the decision-making prowess to succeed in these roles. How, then, do they get them, one might ask? Great question. Most commonly, it's based on *who* they know, not *what* they know. Stating the obvious, I know. Setting nepotism aside, how do I intend on showing said lack of qualifications? You ask such great questions, by the way. This perfectly sets the scene for so much of the content in this book.

Let's begin with utilizing their own words, which were captured as part of a book surveying IT industry leader CIOs (Chief Information Officers). In the book *The Real Business of IT: How CIOs Create and Communicate Value*, they survey CIOs from various large corporations to determine the best practices CIOs should follow to succeed. Completing the surveys and the book led to this groundbreaking conclusion (disclaimer: sarcasm warning). They concluded that *CIOs should take on wider responsibilities within the business outside of IT.* Yes, you read that correctly. I wish I could include a WTF meme within this paragraph. Let's dig into why this philosophy is fundamentally flawed.

I must first preface the forthcoming deconstruction with this important fact. I think most individuals can agree that IT, or tech for short, is how so much of business is done now. All companies need a deep understanding of IT regardless of the industry, from car manufacturers to home appliances, application development, food distributors, etc. Here's the problem: Ask any of these so-called 'executives' what the acronym 'IT' stands for and see how many of them even know. I have a running joke that most would answer 'internet technology' because I've heard this ridiculous answer multiple times before. FYI, it stands for 'information technology,' if you didn't know. And there's no shame in not knowing, unless you're an executive. Yet, so many do not understand even the fundamentals of IT. Blasphemy personified.

With that point made, let's dive into the aforementioned conclusion statement. If CIOs should take on wider responsibilities within the business, what should the business people be doing? What is the purpose of having defined roles with areas of expertise within an organization? What happened to establishing rules of engagement between departments? Needless to say, I was shocked when a book published by the Harvard Press Review came to this conclusion after interviewing CIOs.

Let's utilize an elementary life analogy to show the idiocy in this statement. Imagine if you were the teacher of a second-grade class, and it's time to let the children out for recess. You send them out to the playground, but there's a problem—there's no fence enclosing where they should be playing. You come back out 20 minutes later to find kids all over the place. It takes another 20 or 30 minutes to wrangle up all the kids, including little Timmy, who was making friends with a stray dog—until he bit him. Now Timmy has rabies. Okay, that example is a tad extreme, but I wanted to show how not having a defined set of basic rules can cause far more problems than it solves. This means your staff will become a 'Jack-of-all-trades' instead of a staff of experts. This then, in turn, creates a cyclical loop of partially or incorrectly completed work, creating further weak points, which creates even more problems over time. Let me pose another simple question: If you injure your knee and go to the doctor and determine you need surgery, do you want a general practice doctor to do the procedure or someone who specializes in it? I rest my case.

Allocating work to staff appropriately is one of the chief responsibilities of executives, no pun intended. So, we should just be able to show them the error in their ways, and then they will all change, right? I think we all wish that. The takeaway here is also surprisingly simple. If you're going to engage yourself in the corporate world, understand the executives will always be right (there's also some foreshadowing here for later reasons). Here's why.

In the book *The Personal MBA: Master the Art of Business* by Josh Kaufman, an absolutely terrific book by the way for anyone looking to expand upon their overarching business knowledge, he covers the concept of 'modal bias.' Modal bias is the automatic assumption that your approach is best. In my humble opinion, they should rename this 'executive bias.' To deal with this, follow these tips:

– Accepting the Modal Bias theory will allow you to anticipate executive or corporate BS. Therefore, it saves you the angst of worrying about it in the future. Focus on handling the situation more quickly and efficiently, or determine if you should handle it at all.

– Remind yourself to keep an open mind when approaching these clearly ineffective scenarios and focus instead on enhancing your ability to make wise decisions on how to proceed.

My hope is that you're having as much fun reading said deconstruction as I am putting it together. There are few things I enjoy more in life than slaying massive misconceptions with elementary school facts.

Up next, we will embark upon everyone's favorite topic: know–it–alls. Sarcasm warning again: This topic does not need much of an introduction, as we have all interfaced with them before. Knowing your way around a know-it-all will be all the knowing you'll need for this next reason.

Reason Three: Certain Employees ALWAYS Know It ALL

Trick question time: What kind of books do know-it-alls read? One may think that since they know it all, they've read *all* the books. Far from it. And no, it's not a trick question. Why would they need to read? They already know it ALL. Depending upon the situation, know it all's can actually be quite enjoyable to be around sometimes. They are typically charismatic, well-spoken, humorous, and oftentimes the trickiest trait of all—correct. So, if they are all of these things, why don't people enjoy working with them? This question, in my humble opinion, is the trickiest one of them all. It's embedded in what they seek. Often, they find it. Control. To establish dominion over their work situation. Ironically, control seekers are typically after one true motivation: *safety*. It can be better described as self-preservation. Find a person who is always trying to ensure their own safety, and you find a big problem. How do you proceed? The answers may surprise you.

Given that I have encountered many of these individuals in my career, I find it odd that I am writing about them now. Who wants to think about know-it-alls? I always tried to spend as much time as possible *not* thinking about them. I had to include this chapter, though. You do need to understand how to combat some of these behaviors to have an enjoyable experience in the corporate world. And no, I am not talking about making this a knowledge competition—far from it again. Competition is the bane of existence for all who try. It's not because you will lose; it's because you will end up exhausted. Often, the best battles are the ones that have never been fought.

As you may be noticing, protecting your purpose for taking action is a recurring theme of this book.

Let's jump right into the most useful tips for handling these oh-so-fun individuals, as well as learning in general, shall we?

- When dealing with a know-it-all, only speak up with a combative point of view when you have clear facts and only when absolutely necessary. Make these interactions as impersonal as possible. Do not directly engage with them unless the level of importance is high, i.e., they will make your role or a situation in the company considerably worse if nobody speaks up.

- The key to advancement in life is not knowing more than everyone; it's learning more than everyone. Know-it-alls always have the answer from their own experience. Lifelong learners always seek answers through learning about others' experiences and adaptation. Gather information from various sources whenever possible, especially when the matter of importance is high.

- If you are the smartest person in the room about a topic, you are not in the wrong room; you just become the teacher! If everyone in the room is looking to blame someone else for what's wrong, *then* you're in the wrong room.

The last point here is so impactful. The only thing know-it-alls really know is that they are never at fault. They direct their focus on who to blame instead of how to get better. Meanwhile, there's a trail of failures following their tenured career wherever they go. This story is all too common in the corporate world. Success is not about knowing; it's about executing and achieving results. And not just any results but the most effective results possible. Do what works best in these situations, and you will never have to worry about the noise from the peanut gallery again.

Continuing on the corporate humbling trail, the next reason will get into one of the largest problems these aforementioned reasons create. We are just getting started with the fun. Once we get even more detailed how-to info, you'll begin to see why I am so passionate about the inadequacies of our highly paid corporate leaders.

Reason Four: Senior Management Staff Are Terrible Decision Makers

Sitting in an airport bar firmly supplanted between two diligent corporate workers, I am reminded of how many hours evaded me while working on so many horrid projects. Projects that came by way of terrible decisions—an interrelated, continuous theme, to say the least. How nice must it be to make decisions with virtually no fear of consequences? I kid. There are *always* consequences. Not being well positioned in the realm of understanding what they manage always leaves them more exposed than a demented reality show participant on *Naked and Afraid.*

Let's jump into the best piece of info I've ever heard about making corporate decisions. Speed is always key in making them, but to do so, you need to understand the level of the decision you are making. In an interview, Mr. Jeff Bezos, CEO of Amazon, provided his illustrious position on decisions. Needless to say, it stuck with me as a catalyst for his success. He explained how there are essentially two different types of decisions: two-way door decisions and one-way door decisions. Much like leaving the doors of an airport security clearance section, if you screw up a one-way door decision, the level of rescinding you have to do relates to cheating on your spouse. You *may* be able to repair the damage, and the level of trust moving forward will never be the same.

He further explained that two-way decisions are decisions that can be easily reversed. They should be made quickly, with the intention of failing quickly to identify if said decision will be effective. If not, you can quickly reverse it. One-way decisions—you don't have that luxury. In reality, you can only go through that door once. If you try to go back, the penalty is so much wasted time that you will come through on the reverse side with a slightly grayed hair color. Understanding this importance is something most corporations and their leaders fail to achieve. I am erroneously reminded of this fact by looking back upon one of the many failed business system implementations I've been part of.

The number of one-way door decisions that have been treated as two-way door decisions in the corporate world is immeasurable. They simply don't get it! This is what separates the Jeff Bezos's from the rest of the corporate sheep that seemingly run most of the world's corporations; their terrible decisions are countless, no sleep pun intended. When the rubber meets the road and the chips are in the middle, the decisions you make in those moments define your career. To avoid the fate of so many who go down with these decisions, take the following steps:

- Respectfully call into question the potentially ineffective one-way door decisions from your perspective to your superiors. Document your advice (preferably via email for a quick and easy documentation trail) and track the eventual outcomes.

- If the consequences of the company's ineffective decisions become so egregious that they impact your quality of life, consider switching to a less volatile situation.

Taking note of the types of decisions your corporation makes can be the difference between a healthy and an unhealthy situation. This reason has earned its place of importance amongst all of them. Plus, four is my favorite number, so I always make this number a special one. Make this reason one of the most important in your assessment of corporations you are willing to work for a longer term. It can be the difference between hating your work life or not. And believe me, hating your life due to your job also sucks.

For our next illustrious reason, we will continue our deep dive into the most prevalent reasons for the corporate suck. The minutiae of this world often force them to move at the speed of a sloth, except for when it's most effective for them to pause, of course. I will be rocking my 'Not Fast, Not Furious' sloth t-shirt to start this next chapter.

Section 2 – Systemic and Foundational Challenges

Reason Five: Corporations Move Slowly (Except When They Change)

Corporations sure move slowly, don't they? And when they do, you have to pester them to get moving. Multiple times. Like watching a sloth reach for a tasty leaf or trying to get a teenager out of bed in the morning after they've had a late night. They're slow. Real slow. Kind of like me putting off getting to the points in this chapter. Annoying, isn't it? Now that I am a bit older, my ability to remain patient has improved; however, thinking about this topic already has my blood boiling. To top it off, the most frustrating point of why people in the corporate world are slow is well-known. They know the next steps; they are just difficult to complete. And if it's difficult, you can be sure it will be backburnered quicker than a chore list after a hangover. The reason makes sense, too: Typically, they need rest from carrying the metaphorical helmet, fire-resistant coat, boots, and hose from fighting all the fires needing to be put out.

We will get into the fires in subsequent chapters, but for now, let's focus on the one thing many of us enjoy talking about: speed. And no, I am not talking about taking a prescription drug to assist. Too much speed, and you'll end up not knowing your head from your ass. The epidemic of people abusing those drugs in the corporate world also causes far more harm than good. I'm talking about balanced speed. Speed is so important in business. One of my favorite quotes actually comes from something I thought up. I am certain I'll repeat it, and it's simple to understand: "Opportunity is a moving target." If you wait too long to pounce, you'll miss your shot. Speed is only effective if the work you're doing is also of *high quality*.

Compromise quality, and you will be left with nothing but a trail of messes. Like many things in life, it's a delicate balance.

Now, let's call into question the means of productivity. We all have the means. It starts with accepting the issue. To determine if you have an issue, you can look at your results. Chances are that if you have an issue in this department, you already know. I knew. I have had my fair share of problems in this department. I think it's somewhat normal, actually, and that is one of the few ways I am. Similar to everything else in the corporate world, we cannot control how others act. I feel like this one is systemic, though. I felt it rubbing off on me when I was around it—deprioritizing important tasks because everyone else was. It was haunting. Instead of continuing to adopt those ineffective behaviors, here's what I did to combat them:

- Resistance is your compass. This is by far the best advice I've heard when it comes to the dreaded 'p' word: procrastination. Whenever you get feelings of resistance, it typically means you know you need to get it done. When you begin treating these feelings as a guide for what to do next, you'll be amazed at the results.

- Implement a policy of just getting started and then reward yourself for doing so. Start by taking something difficult and work on it for only five minutes. Just start. Then, at the end of five minutes, reward yourself with a small benefit. For example, if I don't start writing on a day I need to, I do not allow myself to look at the view out of my favorite window in my house. After I write, I get the view. Start with small things. Build momentum. A habit needs to be established before it can be expanded upon. Then, expand.

- Play hours, not results. This is an old poker reference I used to live by back when I played. Set aside time to work, not just to get tasks completed. Sometimes, you'll need to work on something for the greater good of the bigger picture.

- Get rid of soft expectations; set firm ones. Prioritize work appropriately and understand why you need something done and who needs to complete the next steps.

- Escalate on inaction routinely. You may be less liked by some coworkers; however, you will certainly gain respect for wanting things done. If you're not getting what you need from someone, ask their boss to assist. It's amazing how this habit speeds up everything for you when working with others.

- Templatize to create speedier outcomes. We will expand on this in the Process and Automation chapters.

Momentum is a funny thing. Once you get going, it becomes so much easier to maintain. Use this info to help you to get going. I promise you will never regret taking action. Even failures, as many books point out, redirect you to where you are meant to be. Always moving helps you always learn. It's cyclical. The payoffs are nothing short of extraordinary. Cast yourself as the actor in your own story and get moving.

Whew, I need a cigarette after all of that action talk. Up next, we will continue getting into one of my favorite topics of all: how to change. I like change because when you begin taking action, that's what happens. The corporate world, however, has a problem with the partnership of speed *and* change. Knowing why and how to change will be the difference between success and complete and utter failure. Much like the corporate world, failing also sucks. Now that we know more about how to get moving, let's harness it and change how you look at change forever, shall we?

Reason Six: Corporations Try to Change Too Fast

Albert Einstein once said, "The measure of intelligence is the ability to change." I just love kicking off another chapter with an Albe quote! In connection with Reason Four, this one is my most personal bone of contention. I spent *years* of my career dedicated to business system implementations that repetitively failed—dedicating that much time to programs and projects that were not deemed successes taught me more in this space than any other. The info in this chapter will undoubtedly be some of the most important in the book. In the future, flexing your ability to change will prove the measure of your intelligence. That's the kind of flexing I like, bro.

Going back to the two-way door or one-way door decision quote from Jeff Bezos, we learned that the most important decision to make before taking action is to determine what kind of decision you are making. This aligns directly with large-scale change. All large-scale change is a one-way door decision. You *must* get it right the first time because you cannot go back through that door easily or at all. Mid-level corporations of all sizes do not understand this effectiveness fact. Most of the management staff in these organizations want to show how much work they did in shorter-than-expected timeframes instead of getting things right the first time. I have seen millions of dollars flushed down the hypothetical toilet of crappy change.

One time, I even spelled out in an email (so I had proof) to a VP what would happen and how many millions of dollars would be lost if we proceeded with the current business system plan *prior to it being implemented.* They still did not heed the fact-based advice. And two years later, the facts I presented did exactly as I predicted: cost them millions. It's unreal, to say the least. This shows that trying to change too quickly is the equivalent of beginning training for a marathon when you haven't run a mile in years to going out and running 20 miles on your first run. It's merely not going to go well.

In my research, all the most interesting and useful quotes were about change. Per the standard procedure, here are all of the most impactful tips when it comes to managing change:

- In Robert Greene's *48 Laws of Power* book (not my favorite book by any means), I found an extraordinarily useful and accurate quote. Rule 45 states: "Preach the need for change, but never reform too much at once." Make sure to respect the old way of doing things and make any changes to be gentle improvements that occur iteratively over time.

- Always find ways to standardize and optimize during your change endeavors. Always make change *subtle and continual.* I will expand on this point and discuss the how in the Data, Process, and Automation chapters.

- In Rabbi Daniel Lapin's book *Business Secrets from the Bible*, Secret 24, 'Change is Scary and Necessary,' states, "The more that things change, the

> more we need to depend upon the things that won't change." This is another classic example of how to change effectively.

This is how change is done. I cannot understate the importance of change. If you learn to change, you become more intelligent and make work life better for you and the people around you. Now I've got the classic song 'Changes' by Tupac Shakur, AKA my favorite rapper, stuck in my head as we change over to another reason.

Speaking of the next reason, let's continue down the trail of what happens when you change too quickly. Failure. Especially on BIG programs and projects. Why nobody has yet created an F-A-I-L-U-R-E song *à la* the respect song is clearly because failure, like the corporate world, sucks. Let's not fail to move on to the next reason.

Reason Seven: Corporations Fail on Large Programs and Projects

It's true. I am willing to guess you have been part of these failures, too. As we've discussed previously, the reasons for these failures are multifaceted. What's really at the root of the problem, though? This is an amazing question yet again and one we will expand upon in this chapter. In the interview we've covered with Mr. Bezos, he also talks about the criticality of knowing the difference between a decision being reversible or not. At its core, the root of failure on big programs or projects can be attributed to one thing: the lack of understanding of the true nature of the problem. If you do not understand how systemic an issue is, there will be no way to properly address it.

Similar to putting a Band-Aid on a gaping wound or repeatedly putting more oil into a car with a leak in the oil pan, they only address the symptoms of the problem, not the root cause. Being a certified expert in the Information Technology Infrastructure Library (ITIL) framework, I know that issues like these are handled in the problem management procedure. When it's identified that multiple incidents are being opened with a same root cause, a problem case is opened to address the root cause of the issue. While this is a small-scale relation, corporations rarely conduct useful exploration sessions of their own systemic issues. Instead, they contract vendors who implement new business system tools and deem it the forthcoming fix to all their issues. It only causes more issues later on.

Learning to identify the difference between an easily reversible and non-reversible decision can make or break your career. Also, understanding when a company is making poor decisions in these critical moments can give you the foresight required to remain calm and anticipate issues. Therefore, giving you the ability to know what's going to happen ahead of time relieves stress and potentially gives you a call to action to change your own personal situation. Without furthering all that is ado, here are the skills necessary to address these concerns:

- When pondering the seriousness of an issue or problem, define the steps necessary to completely reverse the proposed work. If the steps are simple and not extensively time-consuming, you have a two-way door decision. If the steps are lengthy or touch a much larger systemic issue, chances are you have a one-way door decision.

- Fail quickly on two-way door decisions. Your willingness to fail will bring you the success you need. You either succeed or learn what doesn't work and can try additional methods via tweaking your approach, etc. The only failure is not trying anything with these issues.

- Research one-way door decisions expansively. Understand the true cause of the issue before making *any* changes, and always remember to change as slowly as possible. A company needs to identify internal problems and know what it is trying to achieve first before engaging external resources to assist.

- Consider making changes to your personal circumstances if you see evidence of a one-way door decision that will go awry. In place of complaining, redirect your efforts to determining if dealing with the future problems will be too stressful or not. Preempt future potential mounds of stress whenever possible.

I cannot overemphasize the importance of acquiring as much understanding as possible before embarking upon major programs and projects. The ironic thing is so many of the large ones I have seen approved for scope could have easily been broken down into much smaller, iterative work initiatives that would have succeeded. Failure in these areas is not attributed to a lack of planning but a lack of understanding, as you cannot plan what you don't thoroughly understand.

In the next chapter, I will be taking on arguably the most important detail of this book. The combination of my experience working in both Service Operations and IT has led me to one of the greatest realizations few seem to understand in big business today. It also happens to be one of my all-time favorite topics because it drives so much of what we do in tech. Let's put on our metaphorical diving caps and delve into all the details of managing…DATA! Yes, I know I am a nerd, and laughing at that fact is allowed. On we go!

Reason Eight: Corporations Don't Understand Data

Data. A four-letter word for many. Is it a topic you think you understand well? I thought I did. The countless times I have sat in meetings to discuss it now feel like a blur. I can recall all of the assumptions, misinformation, lack of planning, and an overall sense of annoyance from everyone involved, too. To sum it up in a phrase, data is difficult. There was no real unified approach to managing it at almost every company I have worked for. Having experienced working for a handful of major telecommunications corporations that shall remain nameless, I've had the opportunity to witness many of the different ways it's been managed. They all had one thing in common: They managed data poorly and ineffectively. The most common excuse for how data was poorly managed was usually the same. Like nails on a chalkboard, I can still hear a former VP of mine shriek the phrase, "Our business is so different from everyone else's. It's why we have these challenges." Data, like corporate business, is not as different as everyone thinks. Let's explore some of the factual inaccuracies of this VP's statement.

Unlike the IETF, an internet standards organization, or certifications like the recently retired CCENT and the widely recognized CCNA, most of the important components of IT and business are governed by formalization. Data is different, though. General data standards are something that really doesn't have a dedicated organization to manage them. A Google search will find small organizations and even a small government website throwing in their two cents, but there is no easily locatable major formalized governance initiative. The reason for this is clear: We have dictionaries. As complex as data within business systems can be, the foundation of data literally just comes down to that. Definitions of words. We'll dive into how complex fields can get in a bit, but the foundation will always be the same and start here.

In the book *Everybody Lies* by Seth Stephens-Davidowitz, he states, "Data can be more powerful than our individualized intuition, and can also be weighed too heavily." While his book mostly refers to big and anonymous data captured in Google searches, I believe this quote applies to all data in general. It demonstrates why we need it. Data helps us determine conclusions outside of our initial intuition when proof supports it. It also shows that no data is truly perfect and cannot always be believed. That is why it must be managed in the most efficient and effective ways possible. The importance of *how* data is managed cannot be understated.

The main problem I have seen time and again, both in the corporate world and sometimes even in simple communication with others, is that people simply use words outside of the intended and clearly defined definitions in the dictionary. The chaos this introduces into databases globally cannot be measured. I have witnessed another highly paid VP who did not know the difference between 'quantity' and 'rate' in a simple ERP/finance system. This VP gave instructions to our engineering staff to switch them both internally in our system and on customer-facing invoices. You can imagine how many problems this directive created. I will refrain from utilizing all of the colorful expletives in this paragraph only because I know my conservative Aunt will be reading this book. What a fudging idiot! The excuse for stepping out of definitions of words is almost always the same: “This is proprietary.” What in the actual fudge?

Let’s now dig into the challenge of proprietary information. It’s clearly a necessary evil in business, as it differentiates companies from one another. It was also the excuse both VPs I have referenced in this chapter and so many others have used to validate their positions to justify ineffective major changes. The necessity of utilizing standards across all and, most importantly, other company’s business systems is real, though. Internal and external integrations are a critical component of data management. How do we manage these differences, though? Enter the concept of ‘Data Unification.’

I actually did not know this term or service existed until recently in 2023. I stumbled upon the problem of what I ultimately concluded was ununified data after my years in both service operations *and* IT operations. I saw what few people in corporate business see: the challenges from both sides. Witnessing both the requirements from the business side and the challenges IT had when managing data in their systems became the ‘aha’ moment I had unknowingly sought for years. Data unification helps manage data from both the standards perspective, adopted globally by domain and specialized business systems (CSMs, ERPs, and ITSMs alike) and proprietary information. It helps determine what fields can actually be standardized and which ones need a proprietary designation. Enter the promised detail for most fields that need to be managed and unified:

- Field parameters: Common requirements here include field name, character limitations, total count/length limitations, field type determinations (currency, text box, etc.), informative help description bubbles, etc.

- Field grouping: Information for category, type, and item of a field

- Security restrictions: Determine read/write access for a field.
- Process triggers: Automated process steps carried out by the system once data is populated into a specific field, including integration triggers, etc.

As you can see, the level of detail that needs to be managed in databases is intense. These are just the standard field determinations, too; I am certain additional detail can be dug into that some of you may even be familiar with. I included this detail to show just how important it is to unify these classifications with standards whenever possible. There are a lot of moving parts in databases, and there are also many opportunities to unify and ultimately *simplify* the language of data. Only after a seemingly proprietary field is scrutinized against any and all available standardized fields should it be designated as such. Here are the main points when dealing with data in the corporate world:

- Assess all fields against definitional standards. Point out when the definition of a field or term does not align with the dictionary. Request or complete realignment of any misaligned field names.

- When managing one-off data requests and field requirements, question the hell out of the need for the work and the proper placement of the data. Question why the data is needed, where it originated, and what the outputs are. Understand if another field is already producing the same data in a differentiated format. Remove duplicate data fields as much as possible. Unify and simplify where and whenever possible.

- Preach the need for data unification before undergoing any major system changes or net new system implementations. Aligning business requirements with IT system standards is arguably the most critical part of a successful system implementation. Data requirements need to be at the forefront of any system changes prior to any other requirements gathering.

Putting a bow on this chapter feels like it's ready to be given as a gift to all who manage data. Data done correctly will address both the business *and* technology requirements paradoxically. The downstream problems proper data management prevents are nothing short of immense. Millions of dollars in waste occur each year within corporations unfamiliar with this understanding. Go out, add value whenever you interface with data in the corporate world, and prove your worth exponentially!

Up next is the corporate world topic I am probably most versed in: *processes.* Specifically, process architecture. It's my bona fide area of expertise. What also sucks is that even though I specialized in it for years, I don't have any good jokes on the topic. Maybe someone can tag me in a few over on my IG @thoroughwayne or on Facebook @Wayne Vogel (Author). Maybe the joke here is how little corporate management schmucks actually know about them. Sick burn for the win!

Reason Nine: They Don’t Understand Processes

When looking forward to writing this book, I often wondered how I was going to approach this topic. Probably like you, when I first encountered process documentation, I hated it. I also recall that I was not alone. Most speak harshly of it. Yet when SMEs, managers, directors, or executives spoke about them on conference calls, they all wanted a say in them. It seems like a recurring theme in the corporate world—people wanting control over things they do not understand. Processes are where the rubber meets the road, where work gets done. The details are difficult in some cases. They are almost always monotonous. However, I must admit it’s where the magic happens. I hit my stride in my career after becoming a certified business process architect.

Now, I would never tell you to become a process architect yourself, unless that is a goal of yours. I know it’s not for most. I simply want to discuss the power in them and how to leverage it. Let’s first dismantle the fundamental lack of understanding most seem to have regarding this topic. In the book *The Real Business of IT: How CIOs Create and Communicate Value*, they state the obvious: “CIOs add value through process enhancements.” They also claim that “most if not all IT professionals have foundational knowledge of business service processes.” This gets me mad—really mad! It’s just a factually inaccurate overgeneralization. It also just makes no sense. When are IT pros going to obtain this knowledge unless they have worked in a similar business sector? Most IT pros spend their time in, wait for it…IT, not business services. Yet, this information has been put out by the illustrious Harvard Business Review Press. Another what-in-the-actual-fudge moment to say the least.

These gross generalizations are the lynchpin for them, brushing off the importance of expertise. One of my favorite managers I used to work alongside in my largest and longest-tenured company would always say, 'The devil is in the details.' We all know the truth is sometimes work is difficult and detailed. This is also where the true value sits. Going back to *The Personal MBA* book, they cover the topic of deconstruction, specifically, system deconstruction. Deconstruction is the process of understanding how well a system is running before you can improve it. The simple fact of the matter is that you cannot improve something you do not thoroughly understand. The answers are in the thorough. And now you probably understand my social media name a bit better.

Without going on any more tangents regarding the importance of this topic, let's jump right into the general tips I have:

- When problem-solving or conducting any gap analysis, always start by knowing who *you* are and where you're both effective and ineffective. These are the inputs you will need to know where to either get better or get assistance. Do not vastly overstretch your assumptions or knowledge to get by.

- Understand the power of documentation. Document everything you can, especially when dealing with the most important work. Utilize documentation to break down corporate silos wherever you see them. Expose everything you can before you begin to improve.

- Utilize process documentation as the cornerstone of your business cases for change. Never change anything you do not thoroughly understand.

Having this mentality will add clarity to the way you view corporate business forever. Ironically enough, I thoroughly believe anyone can learn to document processes. It's literally just recording what's happening in the right level of detail. If you do not want to be the documenter (and I'm sure you do not), simply also make sure you consume it as much as you can. Knowing even some of what's going on will help you ask better questions and get to conclusions quicker. My hope is that some of this advocacy will wear off on you and that you will begin to look at processes as the real power instead of control.

On deck is what we get to do with processes once we have them. After you connect the dots on what you can do with processes, you might find your desire to pursue them more…*auto*matic. Segway puns are taking center stage to close out this chapter.

Reason Ten: They Don't Know How to Automate

Automation. How important is it? In a word, very. In a Fortra.com article, they label the five major benefits of automation as:

1. Cost reduction
2. Increased productivity
3. High availability
4. Increased reliability
5. Optimized performance

I'd like to throw in scalability and transparency, to name a couple more. Succinctly put, automation is the way of the future. Eventually, AI will use it to break down most repeatable tasks and create a world where so much of our current work is not completed by us humans. It begs the question, though: Why is the corporate world so bad at it? The answer is in the last reason: *They don't understand processes*. You cannot automate what is not already understood. If you try, you simply create an even larger mess than ever.

Another is a quote from *The Real Business of IT* book, and don't worry, this one is actually true: "IT's Focus = Automation of the enterprise to provide efficiency and scale." This is one of the few spot-on quotes from this book, and this should be a major goal of all businesses, to some extent. How else do you intend on work being completed outside of your employee's mindless all-day social media scrolling? I'm kidding, of course; however, removing non-value-add work off the plates of the employees is yet another benefit of automation. If leveraged correctly, it can profoundly affect employee job satisfaction and substantial return on investment relating to operating costs.

In my time as a lead business process architect, it was my responsibility to locate and complete some of these initiatives. A simplistic example of an effective improvement endeavor was to automate email templates our agents would send out to notify system-monitoring customers of a down system. The emails were templatized; however, some customer details needed to be manually populated each time they were sent. A total of three emails were sent within a 96-hour period. It took agents approximately 15 minutes to populate all three emails. Long story somewhat short: I provided our software development team instructions on how to enhance our proprietary database to automate these alerts through timing and variable-based scripts. This resulted in removing thousands of hours of non-value-add work from the agent's workload and saving the company hundreds of thousands of dollars in employee costs. It all started with documenting the agent's process for sending out these emails. It's neat stuff, in my nerdy opinion.

Imagine just how many repeatable processes can be automated. Many industries, such as automotive manufacturing, machining, and the like, have incorporated many of these practices for years. If the same principles are applied to the business of everyday life, the impacts will be profound. The main takeaways for this reason are as follows:

- Only once you have your process down pat can you generate genuinely scalable requirements for systematized automation. Focus on templatizing and simplifying your processes before moving toward automation.

- Automation will be the number one focus of businesses in the future. Learn to spot automation opportunities in your career where you can initiate improvement requests, resulting in massive ROI and efficiency gains.

- Not all processes can or should be automated. Target the most simplistic and monotonous work with the highest volume first before taking on overly complex automation endeavors.

Take it from me: These are all things I wish I had known much sooner in my career. This type of value-add work was some of the most rewarding I completed. Once you develop the skill of improving efficiencies through process and automation efforts, you begin to see opportunities everywhere. I guess those Jetsons predictions of the future were not that far off, *eh*?

In the next reason, we'll touch upon how important it is to you personally to ensure your efforts go to a worthy cause. I think the science behind the next reason is some of the coolest on the planet when it comes to understanding how we work. Let's keep flowing on through to the next reason.

Section 3 – Day-to-Day Frustrations

Reason Eleven: Corporations Utilize All of Your Flow

Flow. Yes, flow. My hope is you already know what flow is, but if you don't, let's quickly cover the basics. Per Wikipedia: "Flow is the melting together of action and consciousness; the state of finding a balance between a skill and how challenging that task is. It requires a high level of concentration." Basically, it's what athletes refer to as 'being in the zone.' You're completely focused on executing that task, and it's where you do your best work. Let's paraphrase each stage of flow for context:

- Struggle Stage: It's the precursor to flow. Many people struggle to get moving when they move to begin working on something. (In fact, I struggled to start writing this chapter today.) We feel overwhelmed, like completing the task is not doable. Ironically enough, research has shown we need tasks to be a few percentage points stretched above our capabilities to do our best work. Struggling is normal. It's also an indicator that our norepinephrine system is beginning to engage. Arguably, this understanding is the most important stage because if you never get moving, you can never hit flow.

- Release Stage: This is where you take your mind off the task or situation. You simply detach and let go of the outcome. If I need it, I take a quick five- to ten-minute walk and don't think about anything during that time. I detach from any worry or angst and enjoy the movement. This engages our subconscious and allows our brains to chase away the cortisol.

- Flow Stage: In my own fabricated definition, this is the state of work bliss in which you lose track of time and execute in a purely focused way.

- Recovery Stage: Essentially, this is classified as either sleep or relaxation. It can involve a massage, listening to comforting music at the end of the day, intoxicating sleep, etc. It's the time our brain calms down to recuperate for the next day.

Now, why is this important to understand? According to The Flow Research Imperative, we only have 90 to 120 minutes of genuine flow time each day. How we utilize this time is key.

When I began to learn about flow, a lot of my corporate work began to make sense. I compel you to think about that, too. How much of your flow do you utilize on work each day? Is that where you want to be using it? If you allow it, the corporate world will utilize all of your flow each working day. They probably just expect it as you 'doing your best work.' I'm certainly not advocating that we never utilize our flow at work. I am just calling out that it's important to understand what you're committing to each day. Understanding these concepts can help you utilize flow when and where it's most needed—a concept I wish I understood much earlier in my career.

Approaching the challenge of managing your flow is a question everyone has to answer on their own. My advice is to definitely allocate some time to understanding it in depth.

Check out the research done by the author Steven Kotler in this space. He has a fairly newly launched YouTube channel with great content, and he has also written one of my favorite books, *The Art of Impossible.* There's only one key takeaway for this chapter:

- Make sure you understand flow and how to use it wisely.

- Thirty minutes of flow research on YouTube can change the way you do your best work forever.

The next reason continues down another route, which is also close to my heart. Making sure you allocate a certain percentage of your time to learning is of monumental importance if you want to grow as a person. The corporate world does whatever it can to snuff that out for you and benefit them. Okay, okay. That may be a bit too harsh, but I just wanted to make sure you were still paying attention. As tempting as it is, avoid leaving your wingman in this important mission of navigating the corporate world more effectively.

Reason Twelve: Corporations Stifle Your Ability to Learn

In my family's house, a sign hangs at the kitchen entrance, AKA the most visible highly trafficked spot in the house, with our 'Family Rules.' Smack dab in the middle of this sign is the rule: 'We Learn.' As an author and avid reader, it's something I am passionate about. So, when something in my life feels like it's blocking my ability to learn, I take note right away. Without getting into all of the brain science of how much information we can store because, to me, understanding that does not change the outcome, we'll simply focus on why and how this occurs. Working in the corporate world exhausts you with proprietary info, blocking room for transferable skills. Sure, we all need some proprietary info to do our jobs, and still, there's an optimal approach to take when facing these situations.

One of the most important lessons I've learned is to always remain open to learning new things. I'm almost certain we have this in common, as you're reading a book to do exactly that. In all of the business books I read to prep for writing this book, the actual topic of learning was just implied. I know the one skill that has kept me afloat in the competitive corporate world for so long is the one skill I never saw anyone really discuss—learning how to learn. It leads to acting with purposeful intention. It makes motivation occur when you know what to do. I always knew my corporate job silently drained that motivation until I learned how to learn. I believe this is what Mr. Albert Einstein meant when he said, "Know where to find the information and how to use it—that's the secret of success."

Without further ado, here are some of the best ways I made sure to keep up with the trends in an always fast-changing corporate world:

- Set a calendar reminder to review job descriptions for relative jobs in your field once every few months. Note the skills required for each job. If you see a pattern of jobs requiring the same skills, seek those skills out.

- Always request, ask for, and accept free training related to transferable skills in your industry. Your future success hinges on knowing what you need to know to be marketable.

- Document proprietary info as much as possible instead of utilizing valuable memory space for it. Let your notes guide you in completing not-so-common tasks instead of bothering to memorize them.

- As a general rule, work on new transferable skills for fifteen minutes every day. Give yourself breaks from this time as needed during busy times, and never forget to fill slow times with this activity.

These simple and non-time-consuming tricks showed me what to work on to maintain my spot near the top of marketable candidates. I cannot express how handy this came in when I would hit the wall of incompetence at my jobs. Note: It's typically not if but when this occurs at most corporate jobs. Being prepared for those times literally saved me from experiencing countless stressful situations. Combining this skill with others we'll cover in the future reasons became my corporate playbook for success.

Up next is another oh-so-personal bone of contention. Ever heard the expression 'don't tell me, just show me'? Let's analyze what happens when people run the one thing in the corporate world that seemingly never runs out of gas: their mouths.

Reason Thirteen: Corporate Leaders Always Want to Run Their Mouths

I wonder sometimes how they do it. But they always do it. They always find the next thing, the next topic. How do their ears tolerate hearing their own voice so frequently? How does their insecurity survive in the prison that is their mind? I don't get it. Their thirst for control is equivalent to a kangaroo rat's ability to survive in the desert without water. It's just the appearance of control, though. The results and facts of effectiveness will always expose insecurity. Thank goodness.

I am willing to bet we've all been there, stuck in a meeting where the one person who always talks and reiterates every detail *won't quit talking*. They just don't stop. The sound of their voice is like a symphony to their mind. I apologize for the nausea I may have just caused you in that not-so-euphoric recall. One of my absolute favorite sayings can be attributed to the Greek philosopher Epictetus: "We have two ears and one mouth so that we can listen twice as much as we speak." I prefer a bit more varied version, "Listening is twice as important as talking." However, good luck getting this through to these folks over the sound of their own voice.

Corporate leaders can often be compared to mega-church pastors and some politicians with the loads of horseshit that flow from their mouths. The true danger in their verbiage concoction comes from the fact that it usually includes several parts of truth mixed in with their lies. Learning to differentiate between the two is the skill that will set you free from any potential future binds. Coming in hot are these significant tips to help you not only see through them but also make your conversations far more effective for you:

- Learn to quickly identify who the audience is when these people are speaking. Knowing who they are talking to and why will help you to see where they are driving the conversation and for what reason.

- Don't talk about your successes, especially around people you do not know well.

- Always listen to other people who are making money. Say only enough to get them talking about their business ventures, and focus on the how. Never become braggadocious about what you do.

It's ironic. In nearly all the videos I watched on billionaires for this book, there was a common theme: All billionaires do hardly anything in their conversations other than ask questions. The reason for this is clear: They do not need to prove their intelligence to anyone. They simply listen and check the facts for the most effective approaches. I think we could all be okay with being more like billionaires in our business conversations.

Up next is a gripe I have overheard many friends complain about when discussing their jobs. I have alluded to it in previous chapters and finely crafted the art of not giving a fudge about it. I am going to use *all* of my *hands* to smack down these useless time sucks. Without further ado, let's get *meeting* in the next chapter!

Reason Fourteen: Corporations Force Pointless Meetings

This one is an additional reason I added during my time writing the book. It was inspired by overhearing some friends I had over for a UFC fight complaining about the pointless meetings they were required to attend—specifically, all-hands calls. Honestly, I had spaced out on this reason because I subconsciously developed how to handle them years ago. And it's not of higher importance than the other reasons. They want to show the entire company just how great they are. How smart they are. How downright special they are. It's incredulous. People who actually are these things don't need to tell you; they show you.

These company's employees are so great they have to force you to attend. Most companies I worked at took a role call for these types of meetings. Some would even require a comprehension test at the end. Elementary in more ways than one. Then, when you did go, they only provided about three or four things pertinent to you as an employee of the company. They do this because they are trying to validate themselves externally. The ironic thing is that hardly anyone is listening. Since they breed a culture of multitasking, nobody is *really* paying attention anyway. The idiocy is cyclical. If people were proud to work for these companies, they'd attend with pride. Instead, we are left with a poo-show rivaling the lovely smells blowing over from the Midwestern cattle farms.

Covering how to handle these meetings is actually important to developing a quality level of job satisfaction. Here are the tips for this one, predicated on efficiency:

- Protect your ears and your time. Have approved meeting conflicts as much as possible for useless meetings. Find valid reasons to not attend any of them.

- If you are forced to attend, be like them and simply go and do other more important things during the call. Then afterward, either quickly review the sent-out info or ask your manager or one of the illustrious 'favorites' what they felt were the most important points from the call. When they ask you what yours were, agree with them.

- If any executives are reading this book, replace these horrendous calls with a brief and concise email update or one-page slide with only the *need*-to-know facts. Stop dropkicking your staff's morale with fabricated reasoning.

Circumventing these meetings and protecting your time actually has more impact on your job satisfaction than you might otherwise think. Finding creative ways to funnel useless time sucks down the drain always makes corporate life more tolerable. And if you are wondering, no, I did not offer unsolicited advice to my friends. If they want it, they can buy the book too. The smirk I'm sporting right now should hopefully make any of them reading smile, too.

Transitioning to the next reason is yet another one of the very reasons I decided to start my own business. It's a good thing I make the rules on how to proceed here, instead of the corporate world. Some of their rules have about as much common sense as an ostrich has. Let's unbury their heads from the sand and show them why, shall we?

Reason Fifteen: Corporations Have Too Many Rules and Don't Adapt

The number of rules that aren't adapted to current times is never in short supply in the corporate world. They have rules for everything. Rules on how to dress. Rules on what time to be there. Rules on how to speak. Rules on how to act. You get the point. It's one big rule-fest. Sure, some rules are necessary. I am not talking about the necessary ones. I am speaking on the rules that aren't. Nearly every company has them. It's a full-blown authority flex in some cases. They have them because they can and want you to think you can't do anything about it. My stomach is turning just thinking about it. Nothing irks me like a good ol' fashioned power play. They do this because adapting is nothing more than micro-changing, and you now know well how the corporate world feels about change.

Quality business, of any kind, always requires minor adaptations constantly. The reason for this is that our world is constantly adapting. This is not to say that customers don't appreciate consistent, scalable approaches—they do. This is merely a quest to look for opportunities to always improve. Effective rules actually help support this. They help guide everyone to where the best path lies. If there's a point to something, people will respect it. If not, chaos is typically soon to follow. Finding companies with the best rules leads to the best working experiences. Their rules give you a peek into how they are run and, more importantly, who they are. Paying attention to the ones who use them as a benefit is a key indicator of how fit they are for the future.

Let's hop right into the tips for this reason:

- Quietly break rules you dislike with no clear merit to assess what the penalties for breaking them *really* are. Ask for forgiveness, not permission. Find out who's watching and who's not. See who *really* cares. The answers may shock you.

- Favor companies who cultivate creative ways to add benefits by eradicating meritless rules. Rules and lack of adaptation have a direct correlation to your job satisfaction. Find companies led by real leaders who value their people, not their power.

- It goes without saying, but ensure you follow most rules, including all ethical rules, and always maintain a high level of character. The only rules to break should not have any direct, meaningful impact to anything tangible, only perception and subjectivity.

Some rules were indeed meant to be broken. Don't let rules break you. Find enjoyment in your journey. Don't take the power you hold in all of this for granted. Understand that walking the line is an art and not a science. Basically, what I'm trying to say here is, *don't be an ostrich.*

The next reason is somewhat relative to this one. I wish I could bring in Aretha Franklin to sing the intro for this next chapter. Perhaps I could engage AI for assistance, but I think it would just feel off, mostly because this is a book. Don't get disappointed, though, AI. Maybe one day we will have singing business books you create, maybe. I just hope it's past my time.

Section 4 – Cultural and Leadership Issues

Reason Sixteen: Corporate Leaders Don't Respect You (Or Themselves)

The Oxford Languages definitions of respect from Google are as follows:

1. A feeling of deep admiration for someone or something elicited by their abilities, qualities, or achievements

2. Due regard for the feelings, wishes, rights, or traditions of others

I think it's important to point this out upfront because respect has varying definitions. Succinctly put, in the real world, it boils down to showing regard for someone's abilities and worth. At your corporate jobs, did you always think you were respected? Did your superiors respect you? Many times, I found myself in situations where they felt like they were above me due to a title, a relationship with management, or simply because they felt entitled to it. All the while, they rarely give out respect to the people who earn it the most: the workers. Let's explore where this comes from and what we can do about it.

Analyzing what corporate managers look for in individuals they promote gives us some insight into their thought processes. Since many of their claims to fame are working non-stop, getting things done in extremely fast timelines, and 'being the best in the industry,' they tend to look for that in others they promote. It begs a not-so-obvious question, though: Are they self-respecting? It's a question you can ask yourself when looking to climb a corporate ladder. People who validate themselves through their work accomplishments alone typically neglect themselves and everyone around them to *get what they want*. If you have ever worked more than ten hours a day for weeks or even months, you know what I am talking about.

Since we have no control over whether someone respects others or themselves, handling these scenarios has to be about ourselves alone. Here are the best approaches to command respect in the corporate world:

- Take care of yourself physically. Be in shape and dress professionally or appropriately for all scenarios. These are subconscious signs that a person respects themselves.

- Have a presence with your words; be a trained speaker who does not repeat themselves or their points.

- Also, always bring something of value when you speak by knowing your audience. Summate executive messaging to only one or two sentences. In the rare cases where they ask for additional detail, be concise with that as well. Conversely, always include additional details to the relevant audiences who care about it upfront.

- Do not work additional hours when it's not immediately necessary or for elongated periods of time. Respect the other aspects of your life to ensure balance in it.

Some of these things appear quite obvious, but when I look back at my career, I routinely let these things go. Drowning in the sorrow of regret when you look back at certain positions you held is never fun or productive. Maintaining these self-respecting approaches will ensure you become a more effective and, dare I say, happier employee.

Next up will be the topic of ‘superstars.’ Every time I hear that word, I think back to that ridiculous movie where the lady would put her fingers under her armpits when she got nervous and smelled them. Don’t worry; this reason won’t be that gross…or weird. Or will it? There’s only one way to find out—let’s go.

Reason Seventeen: Corporations Cultivate a Superstar Mentality

'Superstar' or 'rockstar' is the interchangeable term business leaders like to utilize when describing their top performers. I greatly dislike it. We will cover staffing guidelines that identify creative ways to motivate your staff. This is not one of them. Stroking someone's ego excessively is nothing more than manipulative and deceitful. It masks the true intent. What is that intent, you might ask? To take advantage of those top performers and not actually provide value of substance, like tangible compensation. This is especially true for the top performers with low self-esteem. Not to overtly mention how it makes the other employees who aren't 'superstars' feel, as it's implied. It's bullshit. They can take their heaps of overabundant praise and put it where the sun don't shine.

The largest sin in cultivating this mentality is that it eventually makes these individuals irreplaceable. In the ITIL framework, they specifically call out removing silos from businesses to ensure repeatable functionality. Superstars are walking silos. They routinely hoard information and important work and, even worse, make things all about them. If you want to cultivate a toxic work environment, just hire a couple of these people. It works great. If you want to optimize and streamline a business, take the advice of a certified IT framework series and *remove* silos and employees who act as such. Yet senior leaders who have never heard of these time-tested and proven concepts cultivate and promote this mentality. Be vigilant in identifying these situations as they can and most likely will become problematic for you down the line if forced to interface with them.

Dealing with this type of culture and individuals like this can be challenging. Make sure you try some of the following tips when handling idiots…*err* I mean 'superstars':

- If you deem an employee irreplaceable, document everything they know. Always leave yourself outs when dealing with these folks, as they will likely turn on you or the company at any point. The owners of any company or business should be the only irreplaceable people.

- Be sure not to directly confront any superstars with anything other than facts. Do not engage in opinion wars with them. Take the high road and remain inquisitive with them about what they *actually* know. Thankfully, these people love to hear the sound of their own voice, so that shouldn't be too difficult.

- Promote a culture of transparency and balance in your business dealings. Do not siphon off work from would-be completers because of visibility or clout. Do not create silos at work, ever.

The best companies do not allow this type of mentality or behavior. Being on a team of high performers doesn't leave room for a superstar—remember that. The difference is clear: team over individual. High performers want everyone around them to succeed and act like it. Superstars or top performers lurk in the shadows, waiting for any high-visibility work they can snatch up for themselves. Maintain a sense of team in your work dealings, and it will benefit you in the long run.

I can promise you I won't be as snarky in the next chapter. I am a golfer, and in golf, when you go to hit your driver long, there's an expression my friends and I enjoy: "Time to let the big dog eat!" I just had to let the big dog eat in this chapter; he was hungry as [insert colorful four-letter word here].

Let's jump over and cover another primary cause of death to understanding: multitasking. I envision a stadium full of booing directly after uttering that word. Let's get focused on the next reason.

Reason Eighteen: Corporations Breed a Culture of Multitasking

If anyone denies this reason, they are lying to everyone, including themselves. We all do it. I am certainly guilty as charged. I thoroughly believe this is the reason meetings suck these days in the corporate world. Nobody is focused on a clear agenda, there are too many complexities floating around to all be managed simultaneously, and they're getting dopamine hits from background direct messages. It's an environment for corporate anarchy. The irony of it all is that people actually pride themselves on being good at it. And what's worse, they can be! I have no doubt seen some folks with this capability. However, as we've covered in specificity during the flow chapter, the cognitive penalty for doing so is massive. The lack of understanding in the work leads to way too many priorities. Then, the chaos of managing too much work leads to multitasking.

I am not alone in this thinking either. Hopping back over to *The Personal MBA* book for a few more key quotes gives this subject all the context it needs. Josh covers both 'monoidealism' and the 'cognitive switching penalty' when addressing this all too important topic. Monoidealism is the state of focusing your attention on one thing solely without conflicts. Cognitive switching penalty is just the scientific term that refers to context switching killing productivity, as pointed out in various studies. I'm not sure about you, but once I learned that multitasking is proven to steal your productivity, I wanted to eradicate it from my working life as much as possible moving forward. I am just one person, though, and I am curious about the thoughts of others here. Maybe it's a topic we can discuss on my Facebook author page or somewhere else.

The way we choose to address this topic in the future will have a direct impact on our work outputs. Choose wisely and follow these principles when confronting these difficult decisions:

- Schedule your day and block out dedicated focus time as much as possible. Remember that focusing on one task at a time is ideal and not always obtainable. We all have to multitask sometimes, but make it a habit to notice when the balance is out of whack and address it swiftly.

- Look into incorporating more flow or meditation time into your day. Set up your computer, phone, and other devices to go into notification dark periods where you only allow certain people to contact you in case of emergencies. Eliminating distractions, especially during flow, will be key to both achieving and maintaining productive work time.

These steps aren't difficult to implement and pay off profusely. Personally, I could never write books the way I do unless I'm in flow. The quality and long-lasting work we accomplish when focused always shows in the end product. Do your best to eliminate multitasking from your day-to-day and watch your productivity increase as a result. It's not a novel concept by any means, but this book would not have been complete without referencing it.

Next up is a reason that is also difficult to navigate. You'll have to trust me on this one; it appears scary on the surface and can be fun, too, once you understand the nature of this beast as well. I tell you no lie, my friends, it's a consciousness raiser. Believe me when I say trustworthiness is a multifaceted topic. Hopefully, I made the next reason clear by my overuse of puns—another dad trait, if I do say so myself. If not, you can bash this book in the review comments on Amazon, Audible, or Apple Books. I'm just trying to have some more fun with the acquisition of knowledge over here!

Reason Nineteen: Corporations Are Not Trustworthy

Much like the intro, I could probably just write 'Duh' and 'The End' and end the chapter. Also, like the intro, I can't, though. Expanding upon this will be important for a number of reasons. I do approach this topic realistically. It's fairly common knowledge that if someone says they never lie, that's true, right? I kid, far from it. In my first book, which, by the way, has nothing to do with business, I speak about my universally adopted, world-renowned formula for building trust (disclaimer: sarcasm warning again). I do believe in it, though, and have yet to hear a better one. It is much simpler than most trust formulas out there, which is why I love it so much. It goes, "Your words match your actions over time." Obviously, several organizations do *not* follow this mantra. The reasons for their lack of trustworthiness are typically dense and not always relative; however, the outcomes are typically the same. They make proclamations they cannot fulfill or guarantee because they don't have the authority to do so. The façade of power is easy to construct with lies.

In the book *Everybody Lies* by Seth Stephens-Davidowitz, when analyzing big financial data gathered on loan defaulting, it was determined, "The more assertive the promise, the more likely they will break it." This can be said for a number of different monetarily based promises of all kinds, up to and including raises or bonuses within the corporate world. The more your company promises and talks about raises, the less likely they are to deliver on them. This notion held especially true for me when accepting promotions or more work. Time and again, you will see this result when up-and-coming employees accept positions with these promises, only to see them never come to fruition. In our most commonly used phrase of the book, it sucks. Understanding how to approach these situations can preempt heaps of disappointment and resentment.

Let's get right into the principles to follow when handling these situations:

- The more assertive the promise, the more likely they will break it. Expect the worst, and hope for the best. Always depend on your understanding that promises of corporate executives rarely fulfill what they claim. Never accept a promotion that doesn't correspond with a raise unless you are prepared to do the work for the experience only. Characteristically, they are just trying to buy themselves more short-term time for your services to delay their self-serving obligations.

- Make sure they include all of the necessary contractual stipulations in your dealings. Also, make sure any of your personal incurred costs are covered contractually whenever possible. Do not do friendly deals, even with known business contacts, ever! Promises aren't real until they are in writing and contractual and *then* fulfilled. Remember, a signed contract does not mean you will get paid; however, it does increase both the chances of that occurring and the legal actions you can take if obligations are not fulfilled.

- To ensure you do lot lose your work templates and files, periodically save backups for all of your work from any company you are employed with on an external drive. Before companies lay you off, they will typically lock you out from downloading any files from your work computer. Not having to recreate these templates can save you hours of time.

Knowing these few approaches can save you more heartache than a country song about divorce. Divorce the idea that people are trustworthy until they have proven their words match their actions over time. If someone burns you, expect they will do it again. Stay equipped with the metaphorical fire extinguisher of this knowledge to ensure the burns don't hit the second or third degree.

Up next is a topic most in the corporate world are familiar with. There's a time and a place for this next reason as well. The first things that come to mind are places, houses, cars, and the like, but not unjustified corporate workers. While this is not my *favorite* topic to discuss, I will do so efficiently it'll seem like it's my favorite reason in the book. My only personal favorites are my readers! You can decide if you think that's also a lie or not. On we go!

Reason Twenty: Corporate Leaders Play Favorites

Isn't this just your favorite topic? Yeah, me neither. I have read lesser valuable business books (that shall remain nameless) that make some pretty wild proclamations. They claim the person who plays best with others always gets the job. Skills don't matter, only being liked. Holy fudging crap, can you believe that? The funny part is, sometimes that's true. Many times, employees of companies are favored for all types of reasons. Maybe they have a similar lifestyle to the boss; they're married with kids, for example. Said boss may lean toward the individual with more personal responsibilities over another equally, or even more egregious, underqualified coworker. It happens often, and this terrible advice encourages it. If you focus solely on relationships and not the value you bring to companies, you may always have a job, but you won't be personally fulfilled with your work.

Now, I would never tell you to not attempt to make some friends at work. It's usually the best strategy, and it makes work more fun, too. Clearly, relationships can also be leveraged to acquire other positions with a company. But to say skills don't matter enters the territory of integrity. If you look at the prototypical corporate staff, these individuals litter the roster. It can be frustrating at first. Then, once you understand the long game, those sentiments fade. Skills are habitually the only thing that will ever get us ahead in life in the long term. It's simply a matter of time before the leaders who hire underqualified individuals become exposed. Ironically enough, though, they do have one skill that matters: puckering up those lips of theirs and getting their noses brown.

The list of guidelines won't shock anyone here. Let's discuss how to approach these situations when engaging with the favorites of the corporate world:

- Expect you will meet the bosses' favorites along your corporate journey. Instead of harnessing bitterness and resentment, replace those feelings with gratitude and the understanding that your skillsets will be rewarded in the long term. Management who unjustifiably plays favorites eventually ends up on the unemployment line more often than not.

- Learn to identify the underqualified favorites and use this info to your advantage. Ask questions to establish how their connection was forged. Learn to make temporary allies when necessary, in place of allowing emotions or judgments to impede your progress on your personalized corporate journey.

- If you are suffering greatly in workload due to a situation like this, look to make changes to your work arrangement as soon as possible. Request work that doesn't require interfacing with them or to be moved to a different team or department before exiting the company altogether. Attempt to be as creative as possible to circumvent these situations prior to them becoming too problematic. Reducing or limiting the amount of interfacing you have with these individuals can greatly increase your role's job satisfaction.

As annoying as these situations can be, you can learn to tolerate them. Sometimes, the greater good of the bigger picture can be key in getting you on your way in your career. It's not a coincidence that the individuals with the skills are usually the ones who win out in the long run. Don't lose sight of that in any moments of frustration caused by favoritism. Develop more skills than Napoleon Dynamite obtaining prom dates or helping Pedro to be elected Class President, and the value of your work will speak for itself.

Up next, we will jump to a reason near and dear to my heart: vendors and partners. Working with a quality one can be a game-changer. Consequently, working with an awful one can be your undoing. Knowing how to navigate this arena will easily be worth its weight in gold the next time you encounter these situations. Okay, maybe not gold, but I had to entice you to keep reading somehow! Marketing was never really my thing.

Section 5 – Operational Inefficiencies

Reason Twenty-One: Corporate Leaders Don't Challenge Vendors or Partners Enough

It's ironic. The concept of a vendor or partner makes sense. Formulate a bond with a company that can execute tasks in their area of expertise when you only need temporary assistance. However, if you let the wrong vendor drive the narrative, you can find yourself in a hole bigger than a moon crater. Many times, I found myself in situations where I knew the vendor in question was exaggerating their efforts to ensure more work for themselves, which is an expense your company will need to flip the bill for. This can happen for a few reasons. The most predominant reason, though, is simply a lack of understanding by either the VP, program manager, or the like in what tasks are really being executed. If no one from the customer company is in the know, the vendor pounces on the opportunity and takes control of the narrative for their own selfish gain. I feel some head nods right now from the ones who know.

To avoid this occurring during the projects you involve them with, you must understand their game. Many of the forthright and quality vendors call out any inadequacies in requirements ahead of a scope of work signing. The opportunistic ones, however, leave as much ambiguity in the scope as possible, just waiting for the right moment in the project to suggest adding a change order that will cost much more than the initial quote. They locate areas of their own expertise your company is not familiar with.

The time that most sticks out in my head is when I saw a vendor tell my then-manager that it took exactly sixteen and a half hours to data dump a simple upload script. I knew it should have taken them approximately five to ten minutes, based on the volume of data. Without boring you with all of the technical details, the time disparity was so great that I knew it was impossible for the engineer not to know how to restart the script or try a different method. They billed overtime hours for their 'efforts.' It was a setup. My former company and boss were being played harder than a recorder in a fifth-grade music class. I was frustrated at the time, and in hindsight, it makes for a good laugh now. It's strange how that works out.

Understanding their ultimate goal of adding work, hours, and fees to the scope mid-project will help you understand their actions. Remaining vigilant and making the most effective decisions during a vendor-involved project can save your company tens, if not hundreds, of thousands of dollars. Let's get right into the tips that will help you deal with preempting these situations from occurring:

- Ensure your initial requirements with any vendor or partner are as specific and detailed as possible. Know exactly what you need them to do and how long it will take. When you are not personally in the know, interface with non-partial experts in the field to determine *actual* completion times instead of leaning on the service provider to answer.

- Question darn near everything you are told by the vendors, especially as it pertains to change orders and additional work requests. They may be trying to leverage recommendations based on their 'best practices' that are wholly unnecessary for your company's situation. Use data from current volumes and scalability predictions to base your decisions on, not capabilities. Need > features, every single time.

- If your company requires a major scope overhaul mid-project, it is better off scrapping it altogether. When dealing with projects or programs with large scopes that are not going according to plan, cut losses as quickly as possible. Show that cost estimates will quickly spiral out of control without detailed scopes of work. Even if you do not have the autonomy to make this decision, suggest and document it. Cite all the areas of weakness and potential leakage.

I cannot understate how important this chapter is when dealing with vendors or partners. If you do not drive the ship, they will drive it for you. And you certainly won't end up where you wanted to go when you get wherever it is you're going. I have seen multiple multi-million-dollar programs run into the ground by vendors. The lack of understanding of data, processes, and, most importantly, change led to large vendor-ran programs that make money disappear quicker than a Federal Reserve recycling party.

In the next illustrious reason, we will discuss what happens after some of these unfounded commitments go awry. I think I already used enough comedic metaphors in this chapter to crack a smile during a super serious business book. Being a poet, too, I am in a poetic mood today. I will try not to overcommit to ineffective comedic attempts here soon. No promises, though!

Reason Twenty-Two: Corporate Leaders Overcommit on Your Behalf

For those of you who already have some experience in the corporate world, you already know there is some truth to this reason, am I right? Some senior leader missed something either in a scope of work or time estimate, and now you're stuck holding the bag. The bag of what, you might ask? I'm trying hard to keep this PG, so don't ask. Many times, instead of just admitting there was a miss, or multiple misses for that matter, when assessing a project, they'd rather just pile the work onto you. When I first wrote the outline for this chapter, the 'why' was because they only care about the bottom line. This is true, and now I think it's also to protect their fragile egos. Who only knows what would happen if they ever had to admit a…don't say it…a mistake! Whatever would they do?

I have another 'fun' question for you. Can you recall any of your current or former employers' mission statements or work values? I can—however, just a few. One of the last companies I worked for actually had a few good ones. The one that sticks out to me the most is the value 'Be your word.' The implication was that if you commit to something, make sure you get it done. This one stuck with me and stands out above all the others. While there's nothing you can do directly when others overcommit on your behalf, you can always make sure you speak up when you think or know the expectations placed on you specifically are unobtainable. It's really the only way to deal with it. Because this will be a shorter chapter, I am just going to jump right into the tips:

- Never overcommit yourself to too much work. Be direct, honest, forthcoming, and realistic in your time estimates, even if it's something your superiors do not want to hear or will anger them. The blame will only shift to you after the fact if you do not.

- Always remember to speak like a lawyer when committing to work that's unclear or has vague requirements. Remaining *committal with contingencies* is how I like to put it. Statements like "I believe this can be done if we do not run into issues" were always valid from my standpoint in these scenarios. Always provide solid reasons for delays that can easily be validated or proven.

- If a company routinely overcommits on your behalf, it may be time to reassess your tenure at said company.

The mistakes I've made in the past in overcommitting to work have left a metaphorical ashy char on my ass. Don't be like a younger me and protect your char-less ass by following this advice. I apologize for any and all inappropriate visuals this paragraph may have given you too.

Up next is the reason why a lot of these budgetary concerns arise. They just don't know how to *sell*. The knowledge in the next chapter will be especially interesting to my non-salespeople out there. Quality salespeople can be the difference in making a place great to work at; I've lived it. Analyzing it was fascinating. It's made me a bit more *je ne sais quoi*, if you know what I mean. 40-something-middle-aged dad swag and sarcasm for the win!

Reason Twenty-Three: Corporations Don't Know How to Sell

Besides never thinking I'd write *je ne sais quoi* in a book, I never thought I'd learn much about sales either. For this book, and out of sheer curiosity, I decided to do just that. As you can guess, in the technology engineering and architecture departments, it wasn't something we focused on that much. However, I always noticed how the top salespeople moved and appreciated what made them great at their craft. Studying greatness in all forms is a hobby of mine. So, after I began to understand what made people great at sales, I knew that most of the companies I worked for were not effective at it—at all. They would start by not understanding the company or the products themselves. Then, what's worse, they would try to sell utilizing the product's features instead of selling how effective salespeople sell. It was typically a mess, and that's why I chose to research and write about it.

We'll start where I learned the most about selling, once again referencing *The Personal MBA*. The most impactful dots I connected in this process occurred when he explained what 'framing' is. I would hear people use this expression all the time, yet I could never find a true definition for what it actually meant. "Just frame it up" is something you hear all the time. After reading the true definition, I began asking other people, including salespeople, what it meant. As you can probably guess, I got a different answer. Every. Single. Time. The fact of the matter is that framing is challenging if you don't know its true definition. If you do, I think it becomes easy. I will share it in the tips and principles at the end of the chapter.

The other principles really focus on developing feelings in your potential customers. As some of you probably know, marketing is the process of generating interest in your product or service, and sales is the art of closing and converting that interest into a sale. In the process of sales, beneficial feelings are only generated in a specific way. And if your sales staff does not understand how to execute this process, they are left fumbling through the features of a product they typically don't understand well. Turnover in sales is high, and this reasoning, along with a lack of confidence, is why. Selling is an art and a science. It's not easy. The best sellers are the ones who know they can generate the right feelings in a client at the right time. They understand the product well enough to know what feelings the solution generates, and then they sell the feelings that having the problem solved creates—very cunning and targeted stuff. I will also cover the best definition of confidence I have ever seen as well.

Without further delay, let's hop into the meat of what makes a great salesperson. The best ones use these principles when selling to their clients to generate the best results:

- As promised, according to *The Personal* MBA, framing is "the act of emphasizing the details that are important while deemphasizing the things that aren't by either minimizing certain facts or leaving them out entirely." The more simplistic way I like to put it when I explain it to others is **the emphasis of certain facts and the deemphasis of other facts**. Mastering this skill has a profound impact on your way of speaking about something you are trying to sell.

- The best sellers convince people they *need* what you sell. Talk about their need. The best sellers hardly focus on describing the product features or selling on price, unless questioned on them. A simple example you routinely see demonstrating this would be a billboard on the highway asking, "Hungry? Food this exit now!" People sometimes buy what they want; they always buy what they NEED.

- The best sellers also sell the result or the feelings the product gives, whatever those feelings may be. They also show know-how of a product with no how. That means they demonstrate knowledge of the product without explaining how it works. They do this by utilizing quick social proof antidotes, AKA customer testimonials, showing how satisfied they are with the product, etc.

- They also always portray an aura of confidence in their appearance and acumen. The real definition of confidence = intense trust in your abilities. The true definition of arrogance = exaggerated abilities. They know they are the best and back it up with evidence. They convince you that you will not find anyone else who can do the things their product does at their price point. The best sellers are always confident and don't fake it. Ever.

Doing research in this area gave me a newfound appreciation for their efforts. Selling is a difficult skill to master, and working for a corporation with great salespeople can be the difference between a positive work experience and a negative one, as I have lived both. When the money is rolling in, times are good. Great salespeople make that happen. A big thank you goes out from me to you if you are one of those people!

Coming to a book near you real soon is our next reason, overcomplicating the simple and oversimplifying the complex. Basically, it's just covering more of what they screw up. I know…that's not sophisticated business talk. Sometimes, I just like to say it how it is. I'll take a page out of our corporate friend's playbook and offer you the opportunity to email complaints over to our email address—CustomerService@nobodyismonitoringthisshit.com—and we'll get to them as soon as possible! Or never, whichever comes first. Curse words don't count if they're in a domain name, right Aunt?

Reason Twenty-Four: Corporate Leaders Overcomplicate the Simple and Oversimplify the Complex

A culmination of sorts over the last several reasons essentially results in this reason. You see it all the time. During the writing of this book, a major IT crash was caused by a company that shall remain nameless. This crash was purely preventable. It affected an estimated 8.5 million systems and shut down major global businesses across the world. I believe this all came from this reason. When you oversimplify the complex and overcomplicate the simple in any work, this is the result. Chaos. The corporate world is not much different from life—results matter. If the results aren't delivering on the needs, call into question why. The end result of how work is mismanaged derives from a mass misunderstanding of what they manage. That is the real reason why companies end up here.

What happens when this occurs is difficult to see on the surface. The resources they are responsible for managing end up misallocated. They have individuals working on various initiatives they do not fully understand. Then, that's when they begin doing all of the things in the other reasons, like changing too quickly, creating useless work, becoming reactive, etc. Touching on resource allocation is key here. When there is a solid strategy, governance, and the like, resources can see the bigger picture they are contributing to and operate as such.

Conversely, when there is no bigger picture, they begin making things up on their own to complete whatever work is assigned. The most dangerous people you will ever interact with in the corporate world are those who *think* they know what they are doing when they clearly don't. The neat thing is that determining if they do is quite simple. Just watch the results they achieve and how the work they're allocated to ends up.

To avoid this problem, there are some steps you can take to ensure you do not fall into the same traps. These guidelines will help you ensure you will not, even if you are in an environment ripe with misunderstanding:

- Always ensure you deeply understand the big asks of your role. Ask clarifying questions about why your assigned work needs to be completed, why it needs to be done in a certain way, and what other work it interacts with. Every process has inputs and outputs; understanding them will help you craft your work in the most useful ways possible.

- To make the biggest strides possible in any corporation, seek out the work nobody wants to do. Most corporate world employees are masters at avoiding the heavy lifting. If you see people avoiding an issue, that's your clue that the work is complex. This type of work is where the 'ahas' come from. 'Ahas' = understanding. Only once you understand the real problems can you begin to solve them.

- Delegate or automate proven simple work as much as possible. What's understood doesn't have to be explained. That's a quality indicator, and work is simple.

If you focus your efforts in a directed and meaningful way around the complexity of work, everything becomes clearer. You begin to understand that all work is linked in some way. Processes are cyclical and interconnected in the corporate world. Some are simple, and some are complex. Developing the skill to differentiate between the two will ultimately become an important ingredient to your success.

Up next are a few reasons I've been so excited to write about. They are some of the major issues I wanted to cover as part of this book. A quick thank you for sticking with reading all of this information so far! Your role as a reader does not go unnoticed or unappreciated. My responsibility as an author will always be to bring my readers the best and most effective approaches to complex topics. Life can be pretty great when we know who's doing what, when, and why. The corporate world has taken a giant step back in this area recently. Let's ship these shits up into shape, shall we? Say that five times fast with no mistakes, and you might just have a career as a rapper in your future! Or let's just head over to the next chapter; the choice is always yours.

Reason Twenty-Five: Corporations Don't Understand Roles and Responsibilities

When I began writing the outline for this book, there were a few reasons I was especially excited about. This is one of them. Companies nowadays do not believe in firm roles and responsibilities anymore. Don't believe me? Pull up a few job descriptions on any of the career websites out there, and you will see. The number of responsibilities they want an individual to cover is immense. It can feel overwhelming just staring at the list. The age-old adage 'Jack of all trades, master of none' pops into mind when thinking about this. However, the clever corporate world has some deeper reasons. The overarching mindset is overestimating their abilities, mostly by way of what they think is inference. However, the ability to infer requires a base knowledge in that skill set, something that is simply not possible to have in many disciplines. This is where the trouble starts.

Again, quoting *The Personal MBA*, he also reviews 'comparative advantage,' which states, "Don't struggle doing things you aren't good at. Work with people who specialize. It's better to capitalize on your strengths rather than sure up your weaknesses." Yet another core fundamental of effective business that is outwardly ignored and openly paraded in public job descriptions. Are you surprised? Thinking everyone can wear interchangeable hats is how shotty work products get into the marketplace and how major mistakes are made. It's also how companies are taken advantage of, and it's the employees of those companies who really suffer. Keep this in mind when reading job descriptions for companies looking to hire on the open market. Chances are numerous underqualified individuals are running these organizations.

When I say the corporate world is clever in these endeavors, I mean it. The reason these underqualified people look for other unqualified people, who typically have the best intentions, I might add, is strategic. If they assign you work in five or six different areas of discipline, it's not a matter of *if* but *when* a mistake is made. Believe it or not, these types of managers sometimes bank on it. They sit back while that employee does most of the work, and by the time a mistake is made on a major project, they can rest easy as they have their scapegoat target acquired. Then, if they so choose, they can utilize this well-intentioned employee as the reason something failed. Oftentimes, the well-intentioned person loses their job, while the management gets away scot-free from any burdens of culpability. I have personally witnessed this occur on more than one occasion. These days, many corporations have become nothing more than a liability minefield.

Fortunately, there are some steps we can take to undermine these tactics. Here are the key guidelines and tips for this chapter; they are some of my personal favorites:

- As soon as your name goes on a task, they can and will blame you for it going awry if it does. Ensure you keep solid boundaries on the things you do not specialize in. Prior to embarking upon these endeavors, call out that you are not versed in the area of expertise and that it's a best effort on your part. Also, learn as much as possible prior to starting the work, preferably from experts.

- To gauge the level of bullshit an individual is capable of, ask a suspected individual a question to which you already know the answer and can prove. Sit back and see what their answer is. This is the best tactic to determine what kind of person you are dealing with. Repeat this tactic for confirmation of their character. The best thing about this is you can simply sit on the information and remain incognito with your intent or choose to respond and call them out at an opportune time. Either is effective in its own way but leans toward remaining quiet in almost all instances.

- Job descriptions are a valuable insight into a company's managerial practices. Know what you're getting yourself into if there are more areas of expertise required than plates on a restaurant menu. Lean toward companies with solid, well-defined roles and areas of expertise and stick to working as an expert yourself whenever possible.

Divulging some of these tips is why I knew I needed to write this book. It's time for you to start tagging bullshitters as such and preempting their attempts at controlling us. Roles and responsibilities are among the most important boundaries we have in business. Disrespecting them is ineffective for everyone.

Up next is another reason that's fun for me—hooray! There's a famous Mike Tyson quote that I and many other authors love that goes, "Everyone has a plan until they get punched in the mouth!" If only our underqualified corporate colleagues actually understood planning, or getting punched in the face. I kid; violence is never okay…unless it's contracted, officiated, and on display for all to see. Then, it's as cool as fudge! I *plan* to drop even more knowledge in this next chapter. Let's get it.

Reason Twenty-Six: Corporations Don't Understand Planning

Plan, yet another four-letter word. This one, though, is more complicated than the others. Many people speak harshly of it. Some believe in it wholeheartedly. Some don't do it at all. Usually, when there are vast polarizing positions, the truth lies somewhere in the middle. To a certain extent, planning is effective. Many less valuable business books preach planning and preparation at length, exaggerating every detail and potential event that could occur. There's just one problem. If you do not have the proper corporate structure based on governance and framework implementation, your plans will never go accordingly. That's a *big* problem—one that Mike Tyson understood implicitly, and one that you can too.

Going back to my favorite business book *The Personal MBA*, Josh makes the following impactful proclamation: Regarding the 'planning fallacy,' he states, "As a rule people suck at planning. Planning is guessing. The value of planning is in mental simulation, the process of having to create the plan itself. **Use plans, but don't depend on them**. Focus on working with Speed and Diligence." So profound. This gives us the basis of how to effectively plan when addressing any situation within business. The corporate world is laced with many complex projects and programs. Understanding where companies are deviating from this definition is a key indicator to determining their level of competency in these areas.

I'm sure some of you have worked on big projects or programs before. This is the chapter where we tie it all together.

I am going to state it bluntly so I get my point across. *Big one-off projects and programs are for idiots.* If you ever see individuals trying to take on these big, transformational plans, they're dumb. Real dumb. Why? There's so much that never goes to plan. They've lost before they started. Big projects and programs create loads of predeceasing tasks. That means many of the project's resources are waiting for someone to complete a task before they can begin theirs. This leaves people in a holding pattern and then thrusts them into hurry-up mode instead of workgroups working concurrently on various priorities. As I have witnessed, it's responsible for massive organizational time waste.

Successful corporations plan less and, as discussed previously, change slowly and iteratively via regular continual improvement initiatives. Not much planning is required when you have a proper interdepartmental structure. One-off projects are far less frequent, much smaller in size, and have a much clearer and obtainable scope. There's little need for predeceasing tasks because an enormous and condensed timeline isn't at stake. Success in business is based on compartmentalizing and simplifying, utilizing scalable and repeatable processes. Remember that. Reduce your project scopes this way, and watch your successes increase.

And now, it's with great pleasure that I am proud to present the guidelines for this chapter to you. Yes, I switched up my plan there with how I flowed into the summary items. I'm still just seeing if you are paying attention. Make planning more effective by knowing the following:

- As a general rule, the bigger the plan, the dumber it is. The relativity to large-scale change is the significant correlation here. Stay away from managing anything massive in scale when it comes to planning. Understand you are most likely being set up to fail before you even begin.

- Plan efficiently and concisely, with only the required level of detail—nothing less or more. Keep any and all project updates to stakeholders simplistic and direct. What's already understood does not have to be explained.

- Preach focus on adding more governance and organizational structure in lieu of large projects or programs. If this is impossible, advocate for compartmentalizing and simplifying any massive scopes into smaller phases. Remove as many interdependencies and predeceasing tasks to enable concurrent work as much as possible.

- Beware of anyone who gives advice saying to plan more or over plans a bunch themselves. This results in a spiral of complexity and leads to getting caught in their own web. Overplanning wastes time because plans almost always change.

I cannot stress enough the level of importance the information this chapter contains. The most effective strategies and approaches are interconnected in the corporate world. Knowing them not only makes work more enjoyable for you but for others around you as well. Wait, what? You mean to tell me that work can be…*fun*? Understanding helps make a lot more things fun at work and in life!

The plan up for the next reason is kind of, well…personal. I hate to say it, but things are going to get pretty darn personal real soon. I feel like I need some type of colored sarcasm text for my non-audio book readers out there. Maybe that's something I can plan for another book. Or maybe that's too personal—who knows? On we go!

Section 6 – Business-Specific Challenges

Reason Twenty-Seven: Corporations Take Business Too Personally

Let's face it: All things in life are sometimes personal. There can certainly be a personal element to business as well. Staking your personal integrity based on business ethics is a positive reflection of that. However, if it's all you have in life, business can quickly become *too* personal. Observing how people would predicate all of their value as human beings through their work became a type of people-watching hobby of mine. I would sometimes sit back and watch them work themselves into a frenzy just to feel accomplished or validated at the end of the day. The irony behind it all is that many times, the harder they tried, the less effective they would become. Followed by a chaser of good intentions, their personal motives became a poisonous elixir to their careers.

All too often, companies make decisions that people begin to take personally. Then, the workers would follow suit and make decisions based on personal feelings instead of understanding the situation better before acting. The truth is this type of behavior has negative ramifications on your career in the long run. If you allow the narrative of the company to alter the way you choose to treat your coworkers or cohorts, the reflections never make their way back to the company, only to you. The contagiousness of negativity abounds in these instances. Avoiding it can be tricky, especially if others peer pressure you to hop on the complain train.

Stand in the face of these temptations by utilizing the following pointers:

- People will always remember how *you* act, not the suboptimal circumstances of the company. Build an armor of integrity and refuse to engage in negative behaviors when experiencing challenging circumstances beyond your control.

- Surround yourself with individuals who focus on executing instead of complaining. Find coworkers with whom you can have meaningful and educating discussions. Do not engage frequently with overemotional or perpetually negative colleagues.

- Routinely monitor the market conditions to determine if your company's negative situation is uniquely exclusive to them or the industry as a whole. Make future determinations based on this information and develop temporary coping mechanisms to fade pockets of challenging times.

Falling into a pit of personal despair when a company's working conditions take a slight nosedive is all too easy. Learning to develop your personal brand regardless of the logo you're currently representing can ease the tension that comes with these trials and tribulations. Never taking business personally is impossible if you care, and taking it too personally will leave you morphed into an ineffective mess. Find the balance between the two, and your focus will shift to what you can control instead of what you can't.

This wonderfully segways into the next topic: They do not really understand the one thing they should—business. The fact is most people in the corporate world do not have an understanding of what it takes to build the very entity they work for. Studying some of the most effective approaches gives you insights into what works and what doesn't. Let's work it on over to the next chapter like a tipsy medieval dad working the dance floor after marrying off the last of his five daughters. The joy of that kind of work can be felt for miles around.

Reason Twenty-Eight: Corporations Don't Really Understand Business

The resounding initial reaction to the title of this chapter may also be 'duh.' Although, the keys to why may surprise you. Even though we live in a world where tech is king, most people still seem to value a world of stability—a world of comfort. Having the same job for years is part of that comfort. The dream of the baby boomer generation was to work for the same company for 30 or even 40 years and then retire. Most of us know now, though, that most businesses don't last forever. In fact, only 52 US companies have been on the Fortune 500 list since 1955. Compare that to the nearly 1.7 million companies that employ over 250 employees in the US today. The data supports this and shows why businesses won't last forever. This means you must get as much as you can as quickly as you can while you can.

Exploring the endless reasons why businesses fail is really what this book is about. Breaking business down to the core of what this means, though, will be the focus of several chapters. I have witnessed countless occasions where corporations will spend money on business that's not real. We will expand on this in the next chapter, but setting the stage here is important. Business is not real until they *pay you.* A signed contract does not mean you will be paid. The main point here is important: Businesses need to hold cash reserves and keep them on hand to ensure they can operate without impediment so bills and employees can be paid. Yet so many companies go month to month on their expenses, just like a college kid fresh out of the mini-van. It's mind-numbing.

This then begs the question, why do businesses last? Analyzing this can undoubtedly become complex. However, it can be summated into a simple set of overarching principles. Think of the largest companies that thrive in just about any market. The first ones that come to my mind without any research are companies like Coca-Cola and Pepsi and car brands that weren't bailed out, like Toyota. People do not buy these brands on price. They simply only need to sell on their brand. If you break down what a brand is, the adjectives of *quality, reputation,* and *expertise* are at the foundation. These companies have fewer complaints as well. The value of providing a reliable experience for their customers is at the forefront of their success.

Let's get right into the core guidelines for how to approach business wherever you are interfacing with it, meaning in the corporate world or on your own:

- Businesses that last forever are a statistical anomaly. Look after yourself first, not the company. Never value a company over *you*. If it's your company, make sure you pay yourself first.

- People do not buy on price; they buy on brand. Always be sure to raise your prices with your growing reputation. Capitalize wherever you can.

- Businesses need to keep excess cash reserves on hand once they are fully operative and profitable. If you witness your company engaging in shady financial practices, plan accordingly in your personal situation or just get the fudge out of there ASAP.

Understanding these initial principles will lay the foundation for even more insight into this theme. Business can be complicated. Breaking it down fundamentally will be crucial to understanding it and not letting it frustrate you in the future. Frankly, I love this shit, and I hope you enjoy it, too. It's neat stuff. And yes, I just cursed for real again—sorry, Aunt!

In the next reason, we will start to break down what I preluded earlier. Getting excited about business you don't have is the equivalent of thinking you will get into a relationship with a supermodel after she likes your 'Beautiful!' comment on her social media page. You're setting yourself up for the disappointment of a lifetime. Breaking down the path to victory here helps create sustainability. Onward we go!

Reason Twenty-Nine: Corporations Get Excited About Business They Don't Yet Have

They do, don't they? You know they do. If water coolers could talk, they would confirm this, too. They would tell us about the sales guy talking about the brand-new Tesla he's going to buy once this customer that's so close to signing actually inks this huge deal. Thank goodness water coolers don't have ears, or we'd need to equip them with commercial-grade earplugs. Why do they do this, though? They think that talking about it will manifest it into reality. How you really manifest business into reality is actually quite simple, and they don't do it. They do not truly understand the customer's problems and, therefore, do not add value to their products and services. Bending over backward to give the customer what they say they want should be a telecasted competition on ESPN's channel, *The Ocho*.

The truth about this mentality is more damaging than it appears on the surface. You might ask, "Why can't someone be excited about a big deal, Wayne?" It's a valid question. The quickest and most relative analogy I can come up with off the top of my head is in sports. There's a reason why most first-time quarterbacks do not win the Super Bowl. I won't bore you with all the stats, and it's a real thing. Experience, or even acting experienced in these situations, is key. It rubs off on your customer. Desperation can be sub-communicated here. If the focus turns to appeasing them instead of helping them, the balance of fit suffers. And they may just take their business to another provider they feel there's a better fit, too.

There are some things you can do to ensure this excitement bug does not bite you when interfacing with any deal. Let's jump in:

- Don't ever be giddy when working with customers. The old phrase 'don't count your chickens before they're hatched' is applicable here. This comes off in your interactions and hurts your chances of achieving your goals.

- Always be pushing for more business; do not live in the past. Living in past successes will sink your future. Wake up every day as if you need the money, even if you have it. Stay motivated and build reserves. You never know when you will need them.

- Always make sure your products and services are value-centric. And work with your product teams to enhance them if they are not. Never depend on one or two customers to keep your business afloat or to meet your quotas. Diversify your business portfolios with as many customers as possible (preferably in varying industries) and expand as much as possible.

I guarantee there are many business owners who wish they understood these things before running their companies. Most times, these lessons are learned the hard way. Robust businesses are built on knowledge and understanding. Mirror them in the way you personally approach all business as well.

Up next will be the continued expansion on this overarching theme of lack of business understanding. The next topic will be exactly that: not understanding types of incoming business. Some of you might also be wondering where the closing comedy is here. I guess you shouldn't get too excited about things you aren't sure are coming. Sorry, I had to troll here; I'll make up for it later. Well, I hope so!

Reason Thirty: Corporations Get Excited About Business They Don't Understand

Now, I want to preface this topic by saying that there are many things I don't understand in this world, too. Fingerpainting, aquariums, and edamame are just a few. However, business is something I think everyone should have a better understanding of, especially if you're in the corporate world. They seem to not, though. Time and again, I have witnessed belief in the principle that revenue equates to success, no matter how you obtain it. While it's definitely a metric for progress, it's certainly not the most important one for longevity. Customers are typically looking for a solution to a problem and even a partner in the space of your expertise. They are looking for the value the solution brings. If you focus on gauging them, throw any potential longevity out the window when they come to realize it.

A basic business philosophy so many have when it comes to generating more of it goes as follows—if someone is prepared to pay you for it, be prepared to get it done. This philosophy is just that, *basic*. Let's break down how this mentality is all too dangerous in the corporate world, or most forms of business for that matter. If there are bad businesses (which I'm sure you can attest to a few), there are certainly bad customers as well. Oftentimes, these 'bad customers' know all too well how to get you into a contract that they knowingly have a provider sign, all with the sole intention of taking advantage of it. Just a few things they like to exclude are lack of volume maximums, specificities around dedicated support, and hyper-aggressive service level agreements, to name a few. If a white-collar service provider does not have a solidified product offering, bad customers come in and create an advantageous one for themselves.

Does this mean you should flat-out say ‘no’ to this business? That is a terrific question. You certainly can; however, there are ramifications for doing so. To high-profile clients, this appears like a lack of capabilities for your organization, which is an image you certainly do not want to portray if you want to do business with them in the future. It’s a fine line, and walking the tightrope of inept appearances is a tricky one. To help with this, follow these guidelines to help manage these situations:

- The quote ‘know yourself to know others’ also applies to your business. To price your services appropriately, ensure you fully understand your business’s capabilities and limitations.

- Never say ‘no’ to business; just ask for more money to execute the services. The harder the work, the more you ask. There are other ways to communicate ‘no,’ so be creative. While it’s unlikely, you never know who will sign a contract with major profit margins to execute services that will ultimately help your business. Let them be the ones to say ‘no’ to your offer instead of questioning your capabilities. If possible, be aware of customer’s reputations in the market as well.

- Corporate mergers and acquisitions in most corporate settings are largely ineffective unless there are massive overlaps within the business models or greater financial reasons for doing so. Always lean toward internal expansion of products and services before purchasing companies to bolster your customer base and increase revenue.

Getting excited about bad business is the equivalent of going on a rollercoaster right after eating a full meal with a funnel cake for dessert. It might be fun to start, but you won't like how it ends. Knowing how to vet your customers is just as important as crafting a valuable offering. It's all intertwined. Understanding these complicated realities will save you more trouble *and* money than you can even imagine.

The next stop downstream after acquiring said customers is dealing with the staff who deliver the services. This is another little fun reason, too. It's also another tricky one, to say the least. It might even be trickier than a Run-DMC rhyme scheme after having some mouth-numbing dental work, but how you handle your staff will unquestionably impact the success of your business overall. Let's keep it moving!

Reason Thirty-One: Corporate Leaders Don't Understand Staff

Hey. Welcome back for another adventurous reason into why the corporate world sucks, with your host, me, Wayne 'The Pain' Vogel. *Okay*…I'll stop writing like a radio host, stop utilizing my childhood nickname, and get on with the book. I just wanted to see if you were still paying attention. Speaking of paying attention, you know who's hard to get the attention of? Corporations, when providing feedback on how to manage their staff. The lack of understanding is deeply enrooted in their inability to recognize what motivates them. The funny thing is money alone will never motivate staff. It may temporarily, but eventually this will not last. If your employees don't like you or enjoy what they do (at least to the point where they don't hate it), you have to pay them a lot of money to deal with it. Worst of all are the useless meetings where they drone on and on about company metrics and specifics that could simply be posted and read on their own if they cared. Instead, they will have someone read every line of this mostly nonapplicable data and spend little time on the outcomes related to the standard employee. I'd bet you know exactly what I'm talking about. Most companies do not know how to treat their staff well.

In prior chapters, we touched on a few different mentalities corporations have when dealing with staff, but we will put a bow on it here. The intertwinement with other staff issues also makes this subject complicated. We will end with how the best companies manage these problems. The most effective companies I have seen and experienced did their best when scaling their teams slowly. They built a small and lean team of the best performers. When they ran into issues with some personnel, they always sided with the loyal employees over the 'rockstars.' If the employee wasn't loyal to management, they would simply let them go. They exited underperformers early, often in cases where it was character-related, not situational. There wasn't much additional determination if the individual was an intellect or overly smart. Loyalty was always valued over those somewhat rare traits.

When teams are smaller, camaraderie can grow organically, even in large corporations with thousands of employees. It's also easier to see if everyone is carrying their weight. Also, when specialized work was needed for a one-off sizeable project, it was much simpler to outsource the work temporarily to vendors. The approach was sensical. Employees either help you make money, or you don't need them. These tough decisions were made utilizing productivity data. If a traditionally productive employee was struggling for whatever reason, they would work with them and find ways to address the issue before taking further action. The best managers would put themselves in their employees' shoes and try their best to treat them how they would like to be treated. It is not exactly a novel concept, yet a difficult one for many to achieve.

If you manage your staff appropriately, it makes everyone's lives better. Since this is such an important topic, there are a few more guidelines here than in most lessons. They are also mostly for people managers but applicable to all business persons as well:

- If you identify a loyal and competent person, find them a job. Source the person over perfectly sourcing the work. Most of the time, people can learn different types of work but not loyalty.

- Staff either make you money or save you time. If not, you do not need them. For managers and business owners, make sure you are routinely reassessing the justification for their cost. Do this at least about once per quarter.

- When outsourcing work, do it as cheaply as possible. Your requirements should be the star of the show. Only outsource work if you cannot do it yourself. If you can hire directly for the work, always choose that preference, as vendors have built-in ways to increase cost.

- Come up with creative ways to motivate your most loyal employees via special perks and shoutouts. Always show gratitude for their contributions as well. In most cases, quality people don't need to work for you; they choose to. Make sure you take care of them in more ways than a paycheck.

- Smart or intelligent workers who still routinely cause problems are simply not worth the hassle. Exit any and all habitual problem creators quickly. Protect your culture and morale like you protect your wallet. They're intertwined more than meets the eye.

- Protect your staff's time from non-value-added work and largely useless meetings. Let your staff have some autonomy regarding what meetings they attend or not. When people feel forced to attend meetings that are not prevalent to them, it dropkicks their morale.

If companies followed these guidelines when dealing with their staff, I have no doubt they would see a measurable improvement in their bottom lines. It's the general staff that makes the real difference in successful companies. If they are treated well, the company does well, period. Crazy, right? As you can probably tell, I am super serial about this one.

Up next, we will discuss one of the favorite things corporate leaders also love to do: overwork you. Since they already know it all, why do they need to do the work? You will learn oh-so-much more if *you* do it, but it has to be their way, of course. The smirk I have writing this is the equivalent to one I had when watching the 2024 presidential debate. There's more bullshit flying around than in Pamplona, Spain's streets during the second week of July, AKA the Running of the Bulls. Bullshit isn't a real cuss word, though, right Aunt? Nice. Good shit.

Section 7 – Philosophical and Big-Picture Reflections

Reason Thirty-Two: Corporations Overwork You

This one is such a precarious position for many, including me. I have proclaimed time and again that I would never let a corporation overwork me, make me work on the weekends, or work extra hours. Next thing you know, I am ordering Chinese food at a quarter till 11 pm after clocking yet another 14-hour day and not eating because "we're under the gun with this project." It's such a difficult subject to manage and address. The simple fact of the matter is companies, specifically senior managers, will take advantage of good-natured employees who want to be successful. All to better their own situation and their personal bottom line. As we have covered, if targets are hit, executive bonuses are paid. And little of that money, if any, ever trickles down to you. Let's get right into how this impacts you and provide guidance on addressing these tricky situations.

One of the first business books I read early in my corporate management career was *Rework* by Jason Fried. To this day, I still believe it's one of the best business books I've ever read, and I highly recommend it. The main point of the book I took away was that work that is done beyond seven to eight hours in a day will likely need to be redone and is not of high quality. As we've discussed in the flow chapter, research has shown it could also be much less time depending on the type of work. Rework is literally just wasted time. This book was also written before the advent of stimulant drugs that will remain nameless to *speed* through these long days. All they really did was speed up the atrocious work, which will need to be reworked at a later date. It's a whole-ass problem I will continue to bring awareness to while addressing all the problems it creates.

Moving on to the effects it has on employees also creates a clear picture. Have you ever seen a person pushing two or three carts full of groceries at the grocery store? The carts are so full that things start falling off the sides, and getting around the store starts becoming a task in and of itself, not to mention the headaches it causes the cashier. Stacking too many projects on an individual has the same effects. Many times, it can be done; however, the inefficiencies it creates leaves the person in shambles afterward. Instead of finding yourself in this position, abide by these principles when approaching your corporate work to ensure you protect yourself from the demise of being overworked:

- Establish clear boundaries for yourself with your manager. Make the time off a medical requirement if you must. Do not allow your work to take over your life—it's never worth it. There may be temporary pockets of time where you flex in more hours to handle a high visibility project; however, most of the time, you need to protect your downtime as it's critical to performing at a high level when you are working.

- Make yourself an important person at work, not indispensable. Always remember that if you do not make yourself indispensable, you become more dispensable. Be prepared by ensuring you are up to date on relevant continuing education and maintain your marketability in case you are, in fact, dispensed.

- If additional hours are required, request additional compensation for the time. This can be done via bonuses, raises, or vacation buyouts. It never hurts to ask, even if the request is rejected. Most of the time, it will be rejected, but on rare occasions, I have been able to receive additional compensation in other ways, such as one-time bonuses and flexible time off during slow times. Always look for and leverage additional opportunities to negotiate much-earned compensation for yourself.

- Please do not take or depend on ADD or ADHD drugs to work extra hours. The side effects are real and simply not worth it.

I cannot overstress the importance of this lesson. Floating in the pool of regret after missing time with your family just so that a project that was not scoped correctly can be finished is substantive. Do not let this happen to you. Set those boundaries and stick to them, even if it costs you your job. Some authors are so passionate about this topic they have written entire books about it. No job is worth breaking these boundaries for, ever. Keep your middle finger in the holster until it's time to use it, and use it aplenty when necessary.

In the next reason, we will continue with more of the managerial and personal-based content this book would not be complete without. If you don't continue reading, I may have to take back the rest of this book and have it self-destruct on you. Let's hope automation never lands in the hands of manipulators, am I right?

Reason Thirty-Three: Some Corporate Leaders and Employees Try to Manipulate You

When starting to work for other companies, I never would have thought this would be the case. In most jobs I have worked, this really wasn't a thing. As I matured in my career, I found that people, especially in senior positions, began to get more personal with me rather quickly. They started to ask more questions that were personal in nature and did not have to do with work. I simply thought that they were trying to be my friend. Boy, was I wrong. Sharing details regarding my personal situation became a source of regret for me. It became evident only after the fact that they were using this information to influence me for their own gain.

A lot of psychology is also at play when working in the corporate world. The prestige of working for certain logos echoes that of attending certain colleges. It becomes a source of validation for their employees. Validation, they hope, will hook you in and make you feel as if you're special. If you look at the relative concepts in areas such as video games, it provides a similar feeling. Video games provide the illusion of accomplishing something others can't and, therefore, create a feeling of superiority, replicating a kill in nature for survival. And the tricky part is that there's some truth to it. Not everyone can accomplish some of these ventures. They hope those feelings become part of your identity. Then, they can start suggesting things a quality company wouldn't—working long hours, traveling more, and frankly, whatever's called for to make them appear successful. All of this is outside of what you originally agreed to and at your expense.

The truth is we all need some validation for doing a good job. Only you can define how much you feel you need. I can tell you, though, time and again, I have seen intelligent individuals get caught up in situations that ultimately made them unhappy in their lives just to chase these feelings. I'd say that high performers are the individuals at the highest risk of this occurring. However, it can happen to anyone. The danger arises when you compromise your values and do things outside your character just to be validated. Sticking to these principles when entering this tricky world will help you navigate it far more efficiently to achieve your goals:

- Do not overshare intimate details about your personal life with anyone at work who is not in your circle of trust. Ideally, that circle should be small and not exist with coworkers unless you know you can trust them. A good rule of thumb to determine if the information is appropriate to share is if you would mind the info being posted publicly on your social media pages.

- Determine why you work before you begin to work. Determine what you want to accomplish as part of your work journey. Question offered feelings of superiority or validation with bottom-line facts of how this impacts you in your real life.

- Establish your values and boundaries with how much time and effort you are willing to give as part of your job and then stick to it. Their lack of understanding or planning on work time estimates does not constitute your emergency.

- Stand strong in the face of any false allegations made against you. Track what you do and ensure you do not allow anyone to claim you did something you did not do for their benefit. Be vigilant when encountering situations where blame is being divvied up, and do not accept any you do not deserve.

These are easier for some than others. It's important to have friendships and bonds at work, given it's at the appropriate level. Professional success isn't just prestige; it's also about getting what you need out of the situation and standing for what you believe is right. If you witness lots of sketchy dealings at a company, take it as one of the reddest of flags, regardless of how you are treated personally in that moment.

Next, we will be jumping into another relative topic: how they take advantage of *you.* And hopefully I shouldn't need to manipulate you to keep reading! Let's get going and take advantage of the info in the next chapter.

Reason Thirty-Four: Corporations Take Advantage of You

Most of us aren't strangers to this. Some form of this you do accept when you sign that employment offer contract and begin working for a corporation. In its simplest terms, you are agreeing to exchange your time for money. Sometimes, that call goes above and beyond the standard expectation. Frankly, it's understandable. Just like in life, things happen in business, too. What's being referred to here in this chapter, though, is when that goes too far—when someone takes advantage of you for their own bottom line or gain so *they* receive more compensation, a higher bonus, or whatever the case may be. It happens more than you think, so protecting yourself in the face of these events is key to ensuring you end up in the best and fairest position for the person you work for most: you.

Personal survey time: Can you recall the amount of times you received a proactive raise not relative to a promotion and higher than the standard one percent or so for cost of living? For me, it was five times in 20 years. Four of those times, I went out and acquired an offer letter from a competing company. The last time was because it was in my employment contract. The point here is that it's rare. Only one of the seven companies I worked for so far in my career actually did this for their staff. They also had a dedicated rewards team to assess the marketplace and ensure their employees were among the highest paid in the industry. Great stuff, but also not realistic for every corporation out there. Most companies do not proactively try to keep their high-performing staff, for good reason. It's expensive. Most corporations merely hope that you'll stay. And many do.

Thankfully, there are a few basic guidelines to follow when working in the corporate world to ensure you aren't fully taken advantage of monetarily. And wouldn't you know it, I've documented them for you right here:

- Never stay at one company for too long—three to five years at the most unless you've received large raises to keep up with the market. Staying at one company too long will price you into a big hole salary-wise. Your skillsets and knowledge may become too proprietary to appeal to other companies as well.

- Analyze market trends utilizing websites, such as payscale.com and the like, to know what range your current salary falls in versus the local market. If requesting a raise without an offer letter, utilize this data.

- Request your employer to put any compensation promises *in writing*. Until it's in writing, it's not real. Verbal agreements cannot be proven. In the US, written agreements can be taken to your state's local division of labor standards, where you can file a claim online for payment. Trust me, when the government gets involved, you'll be surprised how quickly these issues magically get addressed.

These tips have resulted in me being paid far more in my career than I would have otherwise. Not depending on any company to ‘take care of me’ monetarily and deciding to take care of myself turned out to be all things lucrative. Follow this advice and make sure you keep your compensation as optimal as possible for yourself. If you don’t, the corporate world will eat your time for less without a single thought.

In the next reason, we will cover the motives they use to create excuses for how they work. It’s yet another huge pet peeve of mine. Maybe I should have named the book ‘My Pet Peeves of the Corporate World,’ but I don’t think it would’ve been seen as “unique or specialized enough” for them. I’ll just stick to callin’em like I see ’em. Their excuses match the corporate world—they suck!

Reason Thirty-Five: Corporate Leaders Claim Their Business Is 'So Unique' or 'So Specialized'

Question for you: Have you ever heard an executive claim this on a conference call? Instead of explaining what the work actually is, they'll say, "Our work is so specialized that nobody understands it." It's nauseating. Their approach is actually strategic in nature. It allows them to silo off their work or business units altogether, making them irreplaceable to the company. All are working toward their real motive, ensuring their own safety, which is a recurring theme as to why they do these initially confusing things. Trying to ensure your own safety in the corporate world through these kinds of actions actually works for some, so technically, the behavior can be rewarded at times. However, in the long run, it never works. The memories they make when the truth is exposed only ensures the kerosine they drip when walking over every bridge they take will be ignited once they cross it. And people never forget these types—I bet you haven't either. Eventually, they cannot go anywhere else. I've seen it time and again.

Now, every company has some specializations—work where experts take over. That's not what's being covered here. What is being covered, however, is the primary tactics these folks utilize who claim they are specialized but are far from it. Their favorite tactic I have seen over and over again is something I have to cover. This book wouldn't be complete without it. These individuals will routinely make up terms, utilizing grammatically incorrect names. Oftentimes, the definitions of these works mean something different than what they claim. Then, they will claim the work is 'different' or 'complicated,' when in reality, it's usually a solved process they have not been trained in. Similar to the aforementioned VP story where he switched quantity and rate around, they will label fields or data in a way that contradicts the dictionary's definition. It's basically like building their own decoder for the work. Needless to say, it's infuriating.

Once you see what's really being done here, you cannot unsee it. The truth is, these people think they are so smart by creating their own solutions to the company's problems. The problem is that someone much smarter has already solved most of these problems long before they've encountered them. And if you try to tell them, they will say things like, "You don't understand our company." It's madness. Industry standards exist for many reasons. One of them is to avoid situations like this. Romanticizing less effective solutions to already solved processes only hurts the organization in so many ways downstream. When a company needs to transition to a more scalable business system is typically when the poo hits the fan. Everything then becomes exposed, and the rework takes many times longer than if it had been sourced correctly the first time—all because they didn't want to admit they didn't understand the problem systemically in the first place. What in the fudge?

The lessons here are deep. Let's uncover the most effective way to handle these situations so you can identify these people right away.

- Familiarize yourself with and adhere to as many industry-standard practices as possible, as governed by framework organizations and various business system industry leaders. Always assume a process exists for a business practice somewhere and make that your starting point.

- Shed light on every dark corner of your corporation's business processes with the most ambiguity. Document as much as you can and ensure one-offs are streamlined as much as possible. Ensure a process is truly proprietary by verifying that it has not been standardized elsewhere.

- Silos are deconstructed with the dictionary. There are also ways to determine what data is being captured by reviewing the field parameters. Comparing mislabeled field formats to standard fields provides clues into what data could actually relate to in an industry-standard practice.

Sunlight is always the best disinfectant. Shining the light of truth on this type of work will be the key to cleaning up the corporate world. Okay, maybe that statement is a bit melodramatic, but I care about work being done efficiently and effectively. The most intelligent individuals work to live, not live to work. Getting work done in the quickest and most effective ways possible is what I love about work. It helps create the type of balance in our lives that leaves us the most fulfilled at the end of the day.

Thankfully, the next few topics are a bit lighter in detail. They work to uncover some of the biggest lies the corporate world sells us worker bees as well. I mean, retirement does seem kind of great after uncovering all of these inefficiencies. However, we'll uncover how what *seems* best might not always *be* best. In fact, far from it. Let's keep going!

Reason Thirty-Six: Corporations Promote the Retirement Carrot

At first, retirement seems nice. We watched Milton from Office Space pick up a big check off the floor, hit the beach, and order margaritas with no salt. It's the dream life, am I right? It's everything the corporate world sells to us, too. It's that lovely picture embedded in our brains while we put up with yet another unfulfilling day. All to obtain the end goal of sitting on our asses or doing whatever we'd like all day. *Ahh*, maybe someday. As long as they keep that carrot in front of the horse, the carriage keeps moving. All while your best working years slip away, working for them. It works great! For them, too. It seems like a never-ending trap, but is it?

Hoping back to the *Business Secrets of the Bible* book, Secret #38 = Retirement is unhealthy. Frankly, I was a bit shocked by this when I first read it. Retirement has always been a goal of so many. How could it be unhealthy? But then I recalled how unhealthy my father became rapidly after his retirement. His health declined super-fast, and we lost him fairly shortly after retiring from where he spent the last 30 years of his career. He always talked about getting a part-time gig but never did. The truth is that work does keep us healthy. Purpose gives our spirit the energy to keep going. Work is meant to be fulfilling. What does it fill us with? In a word, life. Whether it be for a company, for our family, or even for ourselves, work helps us stay alive. Work is that important.

Therefore, where we choose to work and what we choose to do become among some of the most important decisions we will ever make in our entire life. Understanding that is the key to the growth that will lead you to find more joy in your work. Without further ado, here are the quick hitters on this topic to round out this succinct chapter:

- Remove the goal of retirement from your mind. Understand that work is meant to fulfill us most days. It's not a race to see who can get to the end with the most stuff. Working is some of the most important time we will ever spend in our lives; make the most of it.

- Think about your work from the end. To be blunt, answer this question: What do you want your eulogy to say about your work accomplishments when you die? Then, let that drive you to get it done. Once you establish the path, the true joy then becomes the journey.

This subject may be brief, and it's probably one of the most important chapters in this book. Knowing what you're doing and why switches this mentality greatly. You are here to work for a reason. Many reasons, in fact. Define what you want out of your work, and you will start driving it instead of it driving you.

We've got another pet peeve up next. I know, I know…not another one. If I ever get another pet, I will have to name it 'Peeve.' I'm sure that would inspire some interesting expressions at the dog park when I start yelling that name. I've never been one for normal, and I'm sure you're thinking, 'That poor dog.' I've also never been able to understand this next reason, either. This outro was about as required for this book as higher education is in the corporate world. Hopefully, what I did there wasn't too much of a stretch, but sometimes we authors have to change these things up for fun. Let's move on to an even higher-numbered reason, shall we?

Reason Thirty-Seven: Corporations Require and Then Don't Utilize Higher Education

I am guessing this is already a topic you're fairly familiar with. The tales inside our family minivans and SUVs eventually point to these life decisions, too. Heck, I know I forced them on my kids, who are now a senior in college and a senior in high school, respectively. The latest Census Bureau findings indicate that 23.5% of the US population holds a bachelor's degree as their highest level of education, marking a significant educational milestone. The motivation is clearly so they can be successful in business, whatever they choose to embark upon. Then you have the good ol' corporate world. Most corporate jobs seemingly require higher education. Do you actually use it, though? I've heard the answer to this question more than any in my career. You guessed the responses—most answers are a resounding 'no,' usually in a sardonic tone. It's simple. Like many things in the corporate world, they just want positive appearances without the substance.

No surprise here; I am going to quote *The Personal MBA* for the final time. Many thanks to Josh for creating such a wonderful book on business basics for me to base several of my positions. Regarding higher education, Josh states: "If you are good enough to get in business school, you have enough talent to do well regardless." While I certainly agree with Josh on this topic, there are clear areas where higher education is required. Most C-level executives of prominent high-earning corporations have master's degrees in business. And as you probably know, doctors, lawyers, and the like also require this education. It's a box they require most exceptionally high earners to check. There really is not a 'right' answer here in regards to acquiring a higher education. It's also not a cut-and-dry situation, to say the least. You can do either. The choice is up to you.

The tips in this chapter will also be brief *and* important. Since there's not much we can do about these requirements, it becomes about the decisions we make instead. Here are my guidelines for this illustrious topic:

- Know the industry you want to climb into with *absolute certainty* before blowing excess capital on higher education.

- If you do not have your college degree and do not intend on getting one, you can always focus on acquiring industry standard certifications. Many jobs will overlook the degree requirements in place of these accreditations, and they are a fraction of the cost. In my humble opinion, make applicable certifications a target even if you have your degree as well.

- Learn to read more efficiently and develop your ability to infer. It's a massive skill to have. Learn to look for the knowledge and the main points as quickly as possible. Then, read applicable books to fill your knowledge gaps in areas of business you are not formally educated in.

After reading these tips, you will probably start to understand why I structured this book the way I did. I did so for my fellow speed readers out there, who probably aren't reading this sentence right now. However, a big shout-out to my every-word readers, too, as you get the all-significant context and, *most importantly,* the attempted comedy with the knowledge. Sometimes a smile, contrite or genuine, helps make learning fun. Wherever you are, I hope you're smiling as we end this chapter!

In the next reason, we'll begin our last section by dissecting the strategic decisions and behaviors of executives and corporations. And from my point of view, nothing needs a good ol' fashioned dissection like corporate strategies! Let's strategically move to the next reason.

Section 8 – Strategic Decisions

Reason Thirty-Eight: Executives Steal Your Ideas

As you might guess, this reason is a personal bone of contention for me, too. For this one, we will jump into a bit more ways to handle these situations gracefully instead of with pure and illogical emotion. This reason also makes all the sense because, of course, they are going to steal your ideas if they are not qualified. As a rule, the lower the qualifications of the executive, the more you have to be mindful of how you deal with them.

Identifying these insecurities in executives can be tricky at times. I am certainly not accusing most executives of being unintelligent—far from it. Most of the ones I've engaged with possess plenty of dark triad traits, a psychology reference for some of us human behavior nerds out there. There have also been entire books I've read dedicated to dealing with psychopaths in white-collar office settings. I'm willing to bet you've seen many of these behaviors too—overcommunicating and taking over meetings by trying to show how smart they are, gaslighting people and accusing them of committing to deliverables they never committed to, pointing the finger at others when programs or projects they are running begin to go horribly or downright fail altogether, etc. Frustration typically mounts as they employ these behaviors. And that's what they want. To make you look unhinged while they skate away scott-free from culpability.

Now, I could tell several personal stories of how my ideas were stolen and done so behind my back, only finding out about it later through friends. In a fit of self-control, I'll refrain, though. Just know that it has happened to me more times than I care to admit. Instead of sitting back and taking it, I have decided to do something productive in retort. That's why I have chosen to share the best ways to deal with it. Instead of being hurt, angry, mad, sad, or agitated, learn to do this instead—expect it.

I have read in less useful business books that you should communicate the ideas you personally have the desire to be implemented to your superiors in a way that guides them into thinking it was their idea to implement it. Allegedly, this is done by asking them guiding questions. To me, this is complete and utter useless horseshit. Not to mention manipulative. If I am afraid to share my ideas directly, fearing they won't be implemented or that I won't get credit, I begin to question my future desire to work there. I understand that most people can't just up and quit their jobs immediately, so proceeding with caution is imperative. Here are the most effective ways to handle these situations:

– The less qualified an executive is, the more heedful you have to be of their intentions. Consider refraining from sharing your best ideas altogether and possibly even utilizing them for your own endeavors.

– Create leverage for yourself by utilizing your proposed changes as a carrot to receive a benefit for yourself, such as a raise or promotion. In any proposals for change you create, focus on the desired outcome and not get too far into the how-to.

I'm sure you might be thinking the same thing I did when giving this advice. "My company won't give me a raise or promotion until I have proved myself by doing the job for free." I have also seen business advice suggesting the same thing: to do the job for free first and gain the experience to acquire the actual position elsewhere. As was somewhat previously discussed and in my experience, making this decision is almost always circumstantial. It usually is an effective strategy if you are younger or less tenured in your career. If you are in a position to do the work without the proper compensation, be prepared to never be compensated for it immediately. In my personal experience, I was almost never compensated directly for taking on extra responsibilities. Therefore, I began protecting my time by not taking them unless additional compensation was guaranteed. That worked out well for me in the long run as I became more experienced. Like most things in life, the choice is yours to determine if the juice is worth the squeeze.

In the next reason, we start to dive into some of the deepest contextual explanations for the corporate world sucking. Like, really sucking. The weight of these reasons will no doubt be familiar to you. I cannot wait to continue dissecting all of the lies and uncover the truths of the excuses they use to take advantage of us. Buckle up because the suck is getting ready to head into turbo-charged land as we approach the finish.

Reason Thirty-Nine: Corporations Never Have Budget (Until They Do)

Budget. If you say the word slowly, you can see what they were trying to do: budge it. Fit it in. Estimate your income and expenditures for a set period of time. Every person and company has to do it. Saving money is the name of the game, right? It's ironic. How much of it is done at your…expense? (Dad puns for the win.) Have you ever asked yourself why, though? There were no raises again this year, or only a minimum 'cost of living' raise. "But the company had decent numbers," you think to yourself as you stare at another positive performance review. Where does the money go? I have asked myself this question numerous times, and when I was exposed to the truth, needless to say, I was not surprised.

I'm sure a time or two, you have been promised a raise that did not come to fruition. You worked hard all year, took on additional responsibilities, and completed even more work than originally expected of you. I think most of us who have a decent amount of time in the corporate world have been there. When I climbed the managerial ranks at other companies, I was finally exposed to processes that made me shake my head. I watched as small pools of money were set aside to try and appease the staff, while larger chunks of money went to managerial bonuses when certain budgetary targets were hit. One could argue that the managers put in more hours and, therefore, deserve the extra money. Who's to say what's fair? It's another one of those fun things we just have to accept and figure out how to proceed. That's more easily done with this knowledge, thankfully.

The first thing I've learned in the world of budgets is that it's rare for people to be 'broke.' In fact, hardly anyone is broke; they are just spending money on other things. To get you to spend your money, people will routinely claim it will lead to more money later. Many companies do the same, promising more money later on while you just sit tight and make it through this tough time. Fortunately, I found these effective tips when looking to protect the most important budget of all, *mine*:

- Manage your personal budget stringently. The best way to increase profitability for any business or person is to save the money you already have. Learn to kill FOMO, the 'Fear of Missing Out' on things. Take the shine off of things. Ask yourself if this will matter or if you will use it six months after you purchase it. Can you rent or borrow it? Can you do something else for fun that's cheaper? Scrutinize diligently. Make a 'pros and cons' list for necessary purchases to determine the best approach.

- If a company will not give you a raise after earning one, take your qualifications and shop them on the open job market. Get an offer letter from another company as your personalized business case for a raise. Take it back to your current company's management staff. *Do not resign*; simply present it to them and ask if they can do anything to help make your decision to stay easier. This has earned me multiple raises in jobs where I preferred to stay. Use caution with this tactic, though; make sure it's an offer you are willing to accept. Always be prepared to leave if they do not budge on the raise.

- When looking to switch companies, always assess and know where a company is in terms of its maturity level as an organization. Assuming companies are a good fit without knowing what they're doing is dangerous, even if they are throwing a lot of money at candidates. This is especially true in smaller and fast-growing companies with less than a billion dollars per year in Annual Revenue. Investigate as many data points as you can around a company's average employee tenure, growth pattern, employee size, and revenue numbers. Then, you can plan how to proceed from there.

Taking these precautions will almost certainly ensure you have a more fulfilling and certainly more lucrative career down the line. Clearly, money is the main tool of business. Learning to manage how to approach these matters can be the difference between years of work you have to complete in the corporate world. Looking at it that way always motivates you to make the smaller sacrifices necessary for big benefits later on.

Up next, we will be covering the creation of useless work. When I thought about this topic, I envisioned a person trying to shovel their entire snow-filled driveway with a baseball bat. While I am not certain this is the best or even most accurate analogy, the visual was pretty comedic. A little comedic relief is almost always necessary when dealing with such frustrating topics, don't you agree? And as you can tell by this point in the book, by a little, I mean a lot. On we go!

Reason Forty: Corporate Leaders Create Useless Work

Continuing the theme established by the last few reasons, the outcome of so much of their work is useless. Much of the work they create is based on feelings. And to some extent, it looks productive. They will take a complaint and try to address it with a 'quick win.' By the way, if you know where I can submit this phrase for the all-time greatest oxymoron in the history of the corporate world, please let me know. Their why is simple. They try to justify their positions on utilizing the work they want to do instead of the work they need to do. All to make it look like they are more important than they are in their jobs. Thankfully, there are clues on identifying these time wasters and ways to make yourself far more valuable in less time.

Here's one of my favorite quotes, which I think perfectly applies to this topic: "Understanding your actions *before* you act makes your actions understood." Action taken on initiatives like 'quick wins' are easy to spot as they don't have much understanding behind them. You can clue in on this by measuring the results—the true impact of the work. Value-add work almost always has larger strategic goals fortified behind the actions, backed by evidentiary support, or succinctly put, data. Non-value-add work is usually based on someone's thoughts or judgment about something and clearly lacks attachment to a larger, more understood initiative. A question to clarify the difference between the two will lead you to your answer rather swiftly.

Understanding how to distinguish between useless and useful work will be a major catalyst in not hating your time in the corporate world. Here are the main points to focus on when navigating the rugged terrain of useless work creators:

- Avoid useless work creators as much as possible. Never spend exorbitant amounts of time on work not backed by larger strategic initiatives. Do not work on problems that don't truly exist or don't have the evident warning signs they will exist soon and block scalability.

- The most effective employees and managers partition out work or info that is not relevant to them! Learn to prioritize your work appropriately between value-add and non-value-add work by identifying the difference between the two. Question work justifications when it's not clear what they are. This reduces stress and helps explain why and where you are spending your time.

- Track all of your main initiative work weekly via a simple log. This activity should never take more than 15 minutes a week and should be summarized in quick bullet points. Always have an answer to the question, 'What have you been working on and why?' Make your value clear and track it along with the returns on that work consistently.

Had I known how to do these things much earlier in my career, I would have hated the corporate world a lot less, not to mention avoided a lot of stress. Always make your actions understood. The long-run version of yourself will be so glad you did.

Around the corner in the next reason is another fun topic for me. Becoming more entrepreneurial in my life has shown me just how much the corporate world doesn't value it. Maybe we can start referring to corporate executives as 'entre-manures' just for fun, too. I wonder if they'd see the purpose in that work?

Reason Forty-One: Corporations Aren't Entrepreneurial

This reason didn't hit me until later in my career. Watching companies try and adapt to customer requirements all the time became the biggest bone of contention. Trying to stretch the current capabilities of the products or systems they offered, promising deliverables we had little capability to handle, or just flat-out lying about the benefits of our services were all actions I witnessed. This then created a backend cycle. When all of these were patched together, one-off requirements started causing issues, the promisers were nowhere to be found, and we were always left in reactive mode, trying to clean up the mess. Living in reaction mode is the antithesis of being entrepreneurial. It's why they cannot be it.

If you have ever felt this way, too, you're not alone. I fully believe corporations force you into this mode or way of thinking with them. For the longest time, I did not think that people actually strategized. I thought that work was simply addressing the loudest complainer. What a shame. So much joy can be found in well-thought-out and detailed strategizing. Any failures help you learn new ways of how to adapt your approach. The successes help you experience the true joy of adding genuine value to people's lives. Hitting bottom lines is a necessity in all of our lives, and if you do not find ways to enjoy the process along the way, you are left with nothing but bottom lines.

If this annoys you like it did me, then you might be someone who will benefit from the concept of side hustles.

These tips are not going to be about the corporate world; however, some ways you can escape it are by creating for yourself—a topic I have clearly grown to love over the years. Follow these suggestions if you are ever feeling yourself wanting to become entrepreneurial:

- Create an LLC or an appropriately varying type of company for yourself. Many how-to's are on YouTube or in books dedicated to the subject. There is no better way to learn about real business than to embark upon these endeavors yourself. Start small and go from there.

- Do everything in your business for as cheap as possible. Start your ventures with bare bones. Never invest more than $10,000 of initial capital in a business that hasn't provided any ROI. I cannot stress enough the importance of this one, regardless of anyone's promises or future returns.

- Ask yourself if you truly *need* this purchase to make money. Do not purchase items or services for your business unless you know you need them to make money. If a purchase does not directly make you money, don't buy it. Fancy logos and websites are perfect examples of expenses that are not necessary initially. Wait until you need them to make money to buy them. Reinvest profits with the same mentality for growth and scale.

- Always track write-offs via expenditures to legally limit your income tax liability. An overly simplified example would be if you made $100, but the supplies to make that $100 cost you $50. Do not pay income taxes on $100; instead, pay it on $50 because that's your only profit. A simple expense and income log will suffice for this initially.

The liberation you feel from creating your own business is better than a topless ski session through the Swiss Alps. Okay, you got me; I've never done that before. But, creating your own LLC will help you understand business in a way you can never get from working a corporate job. In my humble opinion, it makes business fun, too! Being in charge of your own business can also help you empathize with other business owners in the corporate world. It can also give you insight into when critical mistakes are being made. The value here is immense.

Up next is the final reason—woohoo! What a journey. Rounding out is a topic I thoroughly enjoy discussing: reactivity. It's the perpetual state most corporate folks live in and probably the most ineffective one, too. Imagine that? I am grateful for that, too; otherwise, I wouldn't have a book to write about business! Gratitude is all around today. Hopefully, you can think of something to be grateful for in your business ventures, too. Let's get into the last of our reasons, shall we?

Reason Forty-Two: Corporations Are Always Reactive

I know I have talked about my pet peeves quite a bit already, and I think this one is the biggest of them all. It's the one thing I know I hated about most of the work in the corporate world. If an important customer would have a problem, you had to drop everything else and hustle over. I understand the need for this. Sometimes, mistakes happen. It's inevitable. However, if you do not learn from your mistakes and make enhancements to what caused them, you become stuck in a cycle of reactivity. Then, if you do get a break in the madness, you rest because you're so burnt out from all of the fire drills. A true work schedule requires a balance between reactive and proactive tasks. If not, the reactivity will destroy your business in hidden ways, all documented and discussed throughout this book.

In ITIL, Continual Service Improvement (CSI for short) is a method to identify and execute opportunities to improve IT processes and services and objectively measure the effects of these efforts over time. It sits over all phases of the framework. In reality, it's the backbone of it. I believe in it implicitly, in all my work, and for one simple reason. It works. Implementing a true programmatic effort in any mid-level or higher corporate organization is imperative to its long-term success.

If you do not, I believe the expression 'death by a thousand paper cuts' is what ends up happening. Or at least that's what my buddy told me when we were having one of our infamous business chats. While I have never seen that cause of death on an autopsy, this is clearly a byproduct of not being proactive enough. When you never improve, the worst happens. You need an overhaul, big changes. And we all know now how those go from that Reason. This helps make sense of the real causes of why so many companies fail. While it can't be attributed to one reason alone, I believe procrastination and reactivity are near the top of the list.

Rounding out our theme of capturing the best guidelines, principles, tips, and tricks, let's get right into the ones for this chapter:

- Always incorporate CSI initiatives into your current workload. At least 20 to 30% of your workload should be dedicated to enhancements from lessons learned. Track them however you can within your current workload, using whatever tracking database your company utilizes. If you do not have the autonomy to deviate from your current workload, speak with your management about implementing this initiative in your workstack. They usually welcome it.

- When you feel stuck in any work situation, locate something small you can do to get you started toward the end goal. It could be as easy as observing what's happening and then writing an email describing the problem. Determine small steps you can take to make progress. Some progress is always better than none. Give it five to ten minutes, and then move on.

Proactive work always ends up starting off small and paying off later in the future. Integrating more proactive work into your workstack will help you and the company in many ways. The best way to solve problems is to proactively avoid their creation. In business, these opportunities are aplenty. If only we could proactively stop the toddlers in our lives from crapping their pants before they're potty trained—that'd be the best proactive work of all time! Maybe AI will create a bathroom brain-reading trigger for kiddos one day; who knows? Avoid the diaper duty of the corporate world and start getting more proactive with your work moving forward.

Conclusion

There you have it: why the corporate world sucks and the secret lessons to success they do not want you to know about. This book was years in the making, and I am especially thankful to my readers for coming along on this journey with me! My goal was to help you learn and understand the truth about all the corporate lies and lessons in business that nobody will teach you so you can quickly scale your career with confidence and avoid additional stress. I know if I had this info before embarking upon my corporate career, the amount of stress and anxiety it would've saved me is immeasurable.

As you move forward in your corporate career, always remember when you're riding on the backs of snakes, you're still slithering. Be cautious, do not get caught up in their ways. Make complaining an opportunity for adaptation and change. Most choose to focus on what they can't do in work situations; ask yourself what you can do. Corporate careers can routinely feel laborious. Suppressing any doubts you have to in order to achieve what you're meant for is an endless struggle intended only for the resilient and confident. Have belief and intense trust in the abilities you have been blessed with; only then can you overcome any doubts, making you who you were truly intended to be. Understand that almost all of your work has a positive impact on society as a whole. Helping create jobs for people that lasted for long periods was always my greatest motivator and honor of my career. Find the reasons that motivate you and understand the corporate world has many benefits as well. And I cannot forget to remind you to attack corporate malevolence with honor and unaffectedness, utilizing the best and most effective tactics to do so.

One of my favorite Bible verses is Proverbs 27:17-20 King James Version (KJV), which goes, "Iron sharpeneth iron; So a man sharpeneth the countenance of his friend." I believe we are all here to teach and better each other. The greatest gift I have been given in my time in the corporate world was my phenomenal mentors. Find them wherever and whenever you can; it's the greatest lesson that can be taught above all the others in this book. I wouldn't be here without them today, and for that, I am forever grateful! Nobody accomplishes anything in this world alone, and having that understanding will help you forge bonds that can last a lifetime.

That's it for me on the corporate world, at least for now. Who knows? Maybe, if this book gets enough traction, I will take *your* reasons and lessons and create a Part Two! I do know one thing: None of this is possible without you, so thank you once again. If you want to continue keeping up with me, you can follow me on all the main socials @Thoroughwayne. Take good care of yourself—you are the most qualified. Until next time!

Bibliography

Kaufman, J. (2010, 2012) *The Personal MBA: Master the Art of Business*. Worldly Wisdom Ventures LLC. https://www.audible.com/pd/The-Personal-MBA-Master-the-Art-of-Business-Audiobook/.

Hunter, R. Westerman, G. (2020) *The Real Business of IT: How CIOs Create and Communicate Value*. Gartner, Inc. and George Westerman Gildan Media. https://www.audible.com/pd/The-Real-Business-of-IT-Audiobook/.

Lex Clips. "Why Jeff Bezos stepped down as Amazon CEO." YouTube Video. Last modified December 14, 2023. https://www.youtube.com/watch?v=vga3PeQlRs4.

Wikipedia. "Flow (psychology)." From Wikipedia, the free encyclopedia. Last Modified August 30, 2024. https://en.wikipedia.org/wiki/Flow_(psychology).

Kotler, S. (2021) *The Art of Impossible: A Peak Performance Primer.* Steven Kotler, HarperCollins Publishers. https://www.audible.com/pd/The-Art-of-Impossible-Audiobook/.

Einstein, Albert. "The measure of intelligence is the ability to change." Goodreads. Accessed September 3, 2024. https://www.goodreads.com/quotes/85475-the-measure-of-intelligence-is-the-ability-to-change.

Einstein, Albert. "Know where to find the information and how to use it. That's the secret to success." Goodreads. Accessed September 3, 2024.

https://www.goodreads.com/quotes/12105870-know-where-to-find-the-information-and-how-to-use#:~:text=%E2%80%9CKnow%20where%20to%20find%20the%20information%20and%20how,That%E2%80%99s%20the%20secret%20to%20success.%E2%80%9D%20%E2%80%95%20Albert%20Einstein.

Epictetus. "We have two ears and one mouth so that we can listen twice as much as we speak." Goodreads. Accessed September 3, 2024. https://www.goodreads.com/quotes/738640-we-have-two-ears-and-one-mouth-so-that-we.

Greene, R. (2000) *48 Laws of Power*. Robert Greene and Joost Elffers, HighBridge, a division of Recorded Books. https://www.audible.com/pd/48-Laws-of-Power-Audiobook/.

Lapin, D. (2014) *Business Secrets of the Bible: Spiritual Success Strategies for Financial Abundance*. Rabbi Daniel Lapin, Audible, Inc. https://www.audible.com/pd/Business-Secrets-from-the-Bible-Audiobook/.

Stephens-Davidowitz, S. Pinker, S. (2017) *Everybody Lies: Big Data, New Data, and What the Internet Can Tell Us About Who We Really Are*. Seth Stephens-Davidowitz, HarperCollins Publishers. https://www.audible.com/pd/Everybody-Lies-Audiobook/.

Fried, J. Heinemeier Hansson, D. (2010) *Rework*. Jason Fried, Random House. https://www.audible.com/pd/Rework-Audiobook/.

Fortra. "5 Benefits of Automation: The Advantages of Automation." Fortra. Accessed September 3, 2024. https://www.fortra.com/resources/guides/automation-advantages-5-benefits-automation.

Perry, M, J. "Only 52 US Companies Have Been on the Fortune 500 Since 1955, Thanks to the 'creative Destruction' That Fuels Economic Prosperity." AEI. Last modified June 3, 2021. https://www.aei.org/carpe-diem/only-52-us-companies-have-been-on-the-fortune-500-since-1955-thanks-to-the-creative-destruction-that-fuels-economic-prosperity-2/.

National University. "73 Bachelor's Degree Statistics and Facts." National University Website Blog. Accessed September 3, 2024. https://www.nu.edu/blog/bachelors-degree-statistics/.

Tyson, Mike. "Everybody has a plan until they get punched in the mouth." Sportskeeda. Accessed September 3, 2024. https://www.sportskeeda.com/mma/news-everybody-plan-get-punched-mouth-how-famous-mike-tyson-quote-originate.

Vanderbloemen, W. Maxwell, J, C. (2023) *Be the Unicorn: 12 Data-Driven Habits That Separate the Best Leaders from the Rest.* William Vanderbloemen, HarperCollins Leadership. https://www.audible.com/pd/Be-the-Unicorn-Audiobook/.

Proverbs 27:17-20 King James Version (KJV) "Iron sharpeneth iron; So a man sharpeneth the countenance of his friend." https://www.google.com/search?client=firefox-b-1-d&q=iron+sharpens+iron+kjv.

Acknowledgments

Teets – Who knew how things would turn out when we met at work? Thanks so much for being the best friend a guy could ever have. Our relatability over the years is unrivaled. Huge thanks to you and Jen for believing in me and always supporting me through all my crazy ventures, WP4L & PKM4L!

Allen – Big Al! You were born to teach and lead teams. Thank you for being the largest source of knowledge in business in my career. Your mentorship and guidance have been beyond instrumental. Thanks for graciously humbling a snot-nosed, self-assured technical engineer back in the day and really teaching me about business! Myself and all the others you have mentored greatly appreciate you.

Bill – Thanks so much for also giving a kid with a small bit of potential a chance. Much like with Allen, some of our business chats leave me in awe of how knowledgeable you are. Thank you, too, for your mentorship, knowledge, and guidance over the years.

To all my favorite coworkers – There are too many to name, but I want to call out a bunch for fun. To Travis E, Tyrone, Ramon, J-Lee, Mix Master Marshal, J-Mart, Joe (Ossy), Kory, Vattman, Mike L, Pat H, J-Hawk, SHopkins, Conrad, John R, Jeff S, Alecia, Ali Bros and Sha, Jarrett, Masooma, Jeanette, Andrew R., Joe P., the other John R., rest of the 'C' Team, Kevin M., Mike D., Dave K., Brian B., J-Nice, Kathleen, Laura, Jen and Mike, Neil E., Denis, Scarlett, Bobby, Jedd, Steve C., Paul McC, Wayne R., Insight Ladies, my homie

Nicholas, Swifty, James A., Moody, Abi, Erin, Paris, The Jens, Pete & Faith, Kris, Anthony T., Kristy, Nicole H., Ian S., Ian H., all the BPA ladies, Don K., Germany Sven, Mark G., Amabel, and my boy and longtime neighbor Eric: You all stick out in my mind as my favorite coworkers over the years! You helped make work fun. Big apologies to any and all I missed!

Steven (Cousin) – Thanks so much for always being my official representation a.k.a. big fancy Lawyer, and designing the amazing cover of the book! Simply put, you're awesome and like a second sibling to me–I'm honored to call you family. Keep up with your personal writing journey too! I'm excited to see where it takes you.

Last but certainly not least, my copyeditor Nicole – Thanks for being my new favorite coworker. Your assistance in bringing my words to clarity is greatly appreciated! And for those of you whom are writing structure sticklers, she definitely recommended that I write the reason numbers above 10 as numerals in line with APA standards. I'm just stubborn and didn't like them like that, go figure! We also shared a good laugh when she had to edit her own acknowledgement statement.

Disclaimer

DISCLAIMER: THE INFORMATION CONTAINED WITHIN THIS BOOK IS FOR EDUCATIONAL PURPOSES ONLY AND IS NOT INTENDED TO BE A SUBSTITUTE FOR CLINICAL CARE OR LEGAL ADVICE. PLEASE CONSULT A HEALTHCARE PROVIDER OR LICENSED ATTORNEY FOR GUIDANCE SPECIFIC TO YOUR CASE OR SITUATION. THIS BOOK DISCUSSES MENTAL HEALTH AND LEGAL ISSUES IN GENERAL AND IS NOT INTENDED TO DIAGNOSE SPECIFIC CONDITIONS OR ADDRESS LEGAL ISSUES IN ANY WAY. USE THIS INFORMATION AT YOUR OWN RISK.

About the Author

Wayne Vogel is a 42-year-old lifelong resident of the greater Denver, Colorado, area. He was a member of several TAG (Talented and Gifted) programs in early childhood and was selected for the IB program (International Baccalaureate) when entering high school. Wayne attended and eventually dropped out of college to exclusively pursue his corporate career following the birth of his first child. He is the father of two children, Madison (21) and Cayden (18), and has been a single father following his divorce in 2015. Some of Wayne's interests and hobbies include hiking, stargazing and sunset chasing, golfing, attending concerts, playing chess, poker & craps, watching sports, and, of course, reading and writing. The highlights of some of his personal and professional achievements include:

- Publishing his first book, *From Brokenhearted to Brave,* in 2023. Along with this book publishing his subsequent additional titles *Flow-With-Me Poetry: Depression Suppression* & *Do You Know the Real You* in 2025.

- Achieving a 'gifted' IQ score of 141, having a Myers-Briggs Type Indicator Personality type of INTJ-A, and currently ranking in the 95+% percentile globally as a chess player on Chess.com.

- Building a career spanning over 20 years in IT, working for Fortune 50 and 500 companies and several other industry leaders. He currently holds or has held several industry-standard certifications in ITIL, PMI, and Cisco Networking.

- Making the money in the only two World Series of Poker (WSOP) bracelet events he has ever entered in 2014, becoming an internationally ranked tournament poker player in Texas Hold'em. He also held the 'Supernova' status on PokerStars when online poker was legal in the US.

- Competing and placing in several World Video Game championships, including becoming the youngest participant to reach the state finals in the 1994 Blockbuster World Video Game at the age of 12.

- Lastly – Don't let all these 'accomplishments' dupe you. Arguably his greatest accomplishment is being considered a fool and a failure to so many! Never listen to anyone telling you that you won't be anything in life. Only YOU get to decide that! I'll always be the little engine that could (thanks Mom)! The work don't lie. Thank God for free will baby, just roll by the haters in your 'Thug Life' shades. Yes, this is really how I talk most of the time. PKM4L too! (Poor Kid Mentality 4 Life) – Kisses and God Bless!

Made in the USA
Monee, IL
12 November 2026

aebc4751-8a0d-4567-996a-93baf4e7733eR01